About the author

Tony Weis is Assistant Professor of
Geography, University of Western Ontario,
Canada. Educated at Wilfrid Laurier and
Queen's Universities, he is particularly
interested in development policy and
practice, and political ecology. He has
published in various journals, including
the *Canadian Journal of Development
Studies*, the *Journal of Agrarian Change*,
Global Environmental Change, and
Capital, Nature, Socialism.

Tony Weis

The global food economy: the battle for the future of farming

Zed Books
LONDON · NEW YORK

Fernwood Publishing
HALIFAX · WINNIPEG

The global food economy: the battle for the future of farming was first published in 2007

Published in Canada by Fernwood Publishing, 32 Oceanvista Lane, Site 2A, Box 5, Black Point, NS B0J 1B0
<www.fernwoodpublishing.ca>

Published in the rest of the world by Zed Books Ltd, 7 Cynthia Street, London N1 9JF, UK and Room 400, 175 Fifth Avenue, New York, NY 10010, USA
<www.zedbooks.co.uk>

Cover designed by Andrew Corbett
Set in Sabon and Gill Sans Heavy by Ewan Smith, London
Index: <ed.emery@thefreeuniversity.net>
Printed and bound in the UK by CLE Print Ltd, St Ives, Cambridgeshire

MIX
From responsible sources
FSC® C019549

Distributed in the USA exclusively by Palgrave Macmillan, a division of St Martin's Press, LLC, 175 Fifth Avenue, New York, NY 10010, USA.

A catalogue record for this book is available from the British Library.
US CIP data are available from the Library of Congress.

Library and Archives Canada Cataloguing in Publication
Weis, Anthony, 1960-
The global food economy : the battle for the future of farming / Anthony Weis.
Includes bibliographical references and index.
  ISBN 978-1-55266-228-1
  1. Agriculture--Economic aspects. 2. Agriculture and state. 3. Sustainable agriculture. I. Title.
  HD1415.W44 2006 338.1 C2006-906312-5

ISBN 978 1 84277 794 7 hb (Zed Books)
ISBN 978 1 84277 795 4 pb (Zed Books)

ISBN 978 1 55266 228 1 pb (Fernwood Publishing)

Contents

Preface

Different worlds of food and farming

I am writing from an immersion in two radically different worlds of food and farming. I am one generation removed from a farm upbringing in southern Ontario, a region that is home to some of the best agricultural land in Canada, and my familial history is a fairly common one for multi-generation European descendants in this country. My grandfather owned an 81-hectare farm, where he raised mostly beef cattle from grains and pasture. When his children saw their future elsewhere, he retired from farming, moved on to a successful business, and sold most of his land while entrusting the last piece as a community park. His sister, my great-aunt, worked a similar-size farm with her husband on which grain production was geared towards feeding a small herd of dairy cattle. She still lives in the same farmhouse but because none of her children or grandchildren farms, the land is mostly leased. Now in her late eighties, she still feeds herself through the year largely out of her impressive kitchen garden, canning, pickling and freezing to sustain her between seasons. A few years ago she commented to me that her district, which was home to about thirty farm families a generation ago, is now mostly operated by one man who leases or has bought most of the land, and his million-dollar machinery and hired workers. While both my grandfather and great-aunt had successful farms and raised happy and healthy families on their land, farm operations of their scale would be economically marginal in Ontario today. The 20 kilometres or so between their two farms is now dotted with large, tin-sided warehouses full of animals, and on a recent summer drive through the area I didn't see a single cow; one *never* sees a pig or chicken raised outside in such agro-industrial landscapes, though occasionally one might see a pig carcass lying at the end of the road waiting for dead-stock pick-up.

Canada is one of the richest and most agriculturally productive nations in the world. And yet, as in most of the temperate world, agricultural bounty does not imply a sizeable or stable farming population. On the contrary, agricultural productivity comes from fewer and fewer and larger and larger farms. Canada's farm population is about one-quarter of what it was at the time of the Second World War and shrinking, and for most young people it is simply unfathomable to think about entering farming today – a situation that is familiar throughout the temperate heart of the global food economy. Typical of industrial farming everywhere, the protracted pressure has been to 'get big or get out' with the high cost of land and equipment and the steady, long-term declines in prices.

In the time that I have been working on this book farmers have taken to their tractors to block a main thoroughfare in the city where I live, and on another occasion farmers from across the province drove a series of highway-blocking tractor caravans to the provincial legislature in Toronto, hoping to draw popular and political attention to the desperate economic situations on the province's remaining family farms. Nearby countryside – some of the best farmland in this country – is increasingly marked with signs that read 'Farmers Feed Cities' and 'Support Our Farmers', in an attempt to draw some broader attention to the struggles of farmers and the future of farming. Also in the time that I have been writing there was a heated conflict between a suburban community sprawling on to southern Canada's largest, but still tragically small First Nations reserve, the Six Nations. The current land of the Six Nations is a fraction of the watershed that had been promised by the British, as their land was eaten into over time by European settler agriculture, towns and cities, including the city where I grew up and the farms of my grandfather and great-aunt. Following not long after the farmers' protests, this conflict set in stark relief the fact that the ongoing social dislocation associated with industrial agriculture here, as in other neo-European landscapes across the temperate world, is layered on top of longer histories of expropriation that have not just vanished.

But the fleeting attention given to these events was only a blip in the longer-term unconsciousness about food, agriculture and how history lives in our landscapes, which is certainly not unique to Canada. And there are compelling reasons why most citizens of industrialized and fast-industrializing countries might give little thought to their food; the volume and apparent diversity could well seem endless. For the world's affluent countries and classes, a global palate is at our doorstep as never before. For instance, the average US and Canadian supermarket has 12,000 items for sale, a range scarcely imaginable a generation or two earlier. The largest volume of items would invariably be drawn from grain and livestock products, processed and combined in various ways, but an ordinary supermarket might also include tomatoes from Mexico, grapes from Chile, mangoes from Brazil, cut flowers from Colombia, shrimp from Thailand, cheese from

New Zealand and oranges from South Africa, along with processed goods with derivatives sourced from far away.

In addition to the supersized supermarkets, in the mid-sized city where I live I can walk downtown and eat at a British, Chinese, East African, Hungarian, Indian, Italian, Japanese, Greek, Latino American, Lebanese, Mexican, Moroccan, Thai or Vietnamese restaurant, and an array of other eclectic, independent eateries, the sort of cosmopolitanism which is surely one of the great benefits of globalization, however much this sort of diversity conceals the standardization and a host of problems in the agricultural system supplying the food. On that same short walk, I would also see a McDonald's, Burger King, Subway, KFC, Starbucks and Quiznos, and a number of other fast-food chains.

But I have also had a profoundly different immersion for some time. Since 1997, my research has focused on the problems and prospects of small farmers in the developing world in the context of very unequal landscapes, debt and structural adjustment, and the changes induced by trade liberalization. Much of my empirical grounding has come in the Caribbean, where agricultural landscapes still bear vivid testament to both the violence of slavery and colonialism and the resilience of the peasantry that emerged after emancipation on the margins of plantations as, outside of Cuba, those landscapes have never faced serious land reform. Thus, struggles of small farmers have extensive colonial roots, and these have been exacerbated in recent decades by adjustment and liberalization which have together driven a dramatic 'de-peasantization' and made the region an extreme case of food import dependence. Coming in the context of job- and production-scarce economies that are now kept afloat by tourism, migration and remittances, Caribbean de-peasantization is entwined with a range of other economic and social problems and illustrates some of the pressures and outcomes of deepening integration within the global food economy in striking ways (Weis 2006, 2004a, 2003).

Aim and outline of this book

The ongoing battle for the future of farming is a momentous one, and the main aim of this book is to examine in a concise and accessible way the major contemporary dynamics, problems and inequities of the global food economy, how this trajectory was forged and is being entrenched, and ultimately different scales for anti-systemic action. All this is guided by the basic assumption that illusions of inevitability need to be unsettled and replaced with an understanding of alternatives as both urgent and possible. This will, it is hoped, help farmers, activists, students and people generally concerned about the health of their food, its producers, the environment and other species situate diverse local contexts within a bigger picture and ultimately help invigorate action.

In the introduction to *Hungry for Profit*, an excellent collection of essays on the political economy of food, the editors rightly insist that 'there can be no longer any doubt today … that we are in the midst of an unusually rapid change in all aspects of the world's agriculture-food system' (Magdoff et al. 2000: 9). And whether food economies and agricultural systems are examined in Canada or the Caribbean, Sahelian Africa or monsoon Asia, the context for understanding the impacts of this rapid contemporary change has inescapably widening dimensions. To recognize this is not to fall prey to a global determinism; as the preceding examples suggest, understanding agrarian change anywhere starts with localized historical geographies. It is, however, to draw attention to the fact that farmers everywhere are facing destabilizing competition as markets become integrated and, as Chapter 1 makes clear, there are very uneven outcomes both *between* and *within* nations.

Chapter 1 frames the global food economy in terms of both its long-term instability and the more immediate economic, social and environmental manifestations of these systematic contradictions, highlighting the major tendencies through measurable trends. It begins by setting out the broad, uneven contours of agrarian change in terms of the simultaneously homogenizing yet wildly imbalanced character of consumption and production, which are being driven by the increasing control of agricultural transnational corporations (agro-TNCs) throughout the global food economy and are underpinned by the tremendous productivity of the industrial grain–livestock complex in the temperate world. The discussion of food consumption patterns draws attention to the revolutionary changes and convergence in human diets on a planetary scale and to the distributive inequalities which are marked by the poles of obesity and hunger. Though the nation-state is too coarse a scale to unpack the imbalances of the global food economy, the imbalances in production are viewed through a snapshot of global agricultural trade patterns which provides an important if incomplete illumination into the understanding of the heated, conflictive politics surrounding its institutionalization via international trade agreements, the subject of Chapter 4.

Attention then turns to the social and environmental impacts and instability of the global food economy on its current trajectory, which are essential to understanding what is at stake in the battle for the future of farming. The twin social convulsions of de-peasantization and rapid urbanization in the developing world are widely outpacing the development of urban (and especially formal-sector) jobs and basic amenities, with the proliferation of slums being the great urban materialization of rural poverty, land hunger and dislocation. The environmental burden associated with the global food economy is discussed first in terms of the toxic burden associated with the industrialization of agriculture, before the lens is then widened to situate the role of the global food economy within the epochal environmental crises of biodiversity loss and climate change.

Chapters 2 and 3 explore the modern foundations of the global food economy, focusing on the great contradiction between the world's heavily industrialized agricultural economies which dominate global markets and prices and the labour-intensive agricultural economies in which the vast majority of the world's farm population resides. The origins of the contemporary global food economy could be traced back through a series of revolutionary changes, which once took shape over the course of millennia, then over centuries, and which are now compressed into mere decades. The intent here is not to make sense of food and farming within the *longue durée* of millennia and centuries (see, e.g. Mazoyer and Roudart 2006; Fernández-Armesto 2002; Sauer 1972 [1952], 1952) but instead to focus attention on the warp speed of agrarian change in the twentieth century, particularly the transformations occurring since the end of the Second World War, which is most critical for understanding the nature and institutional fortification of the global food economy.

Chapter 2 examines the basic political, economic, social, ecological and technological revolutions at the root of the industrial grain–livestock complexes in the temperate world, from the English enclosures to development of high-input extensive monocultures and factory farming. The intent is to explain how the persistent surpluses and subsidies came to be and the forces lionizing scale and driving the radical ecological simplification, polarization of production and concentration of power in massive corporations, with a focus on the archetype of corporate-industrial agriculture, the USA. One of the key objectives is to understand the basis of today's export imperatives and contentious subsidy regimes, which arose in response to the upsurge in productivity and the instability of the emerging industrial model and played a major role in shaping the global food economy and forging the deep import dependencies in parts of the developing world. Another major objective is to appreciate how the growth and consolidation of agro-TNCs has contributed to simplify-

5

ing and standardizing diverse agro-ecologies and farm animal lives while making the ways that food moves from farms to tables become ever more complex and opaque, with production and consumption integrated through vast distribution, processing and retail systems. Appreciating the interests of agro-TNCs and their position as the dominant actors at the heart of the global food economy is central to understanding how they have sought (and largely succeeded so far) to institutionalize the trajectory of the system through multilateral trade agreements.

Chapter 3 looks at the major dimensions of agrarian change in the developing world from the waves of decolonization following the end of the Second World War into the era of debt, structural adjustment and uneven industrialization. It begins with a brief review of one key element of the colonial inheritance – deeply inequitable distributions of land – before addressing in more detail the struggles to challenge, reshape and defend these inequalities after the formal yoke of colonialism was overcome, struggles that remain a crucial aspect of agricultural problems across much of the developing world to this day. A broadly defined three-dimensional framework is then employed to help navigate the diverse terrain of agricultural change in the developing world. The first group of countries examined are those where agriculture makes the greatest relative contribution to GDP and employment in the world and yet which became entrapped in deep, seemingly intractable food import dependence and suffer from the world's most severe food insecurity; this group includes most of the world's Least Developed Countries (LDCs). The second group of countries are those which successfully pursued food self-sufficiency amid rapid population growth by coupling very different productive transformations with similarly protective agricultural trade policies. The central nations considered here are China and India, emerging industrial powerhouses which together contain two-fifths of humanity and have been relative non-factors in global agricultural trade, which make them prime targets for market expansion by agro-TNCs. The third group of developing countries are those with highly competitive agro-export platforms which have come to occupy a prominent place in the global food economy, with these agro-export sectors having tended to solidify or extend extant inequalities in rural landscapes. The differences between the latter two groups, as will be seen in Chapter 4, have constituted a major barrier to constructively challenging how the USA and the EU were designing the World Trade Organization (WTO) in the interests of agro-TNCs. The chapter concludes by examining how structural adjustment reconfigured agricultural policies and compressed state sovereignty, an essential precursor to the institutionalization of trade liberalization in the WTO.

While national lines help to give structure to Chapters 2 and 3, this does not mean that global food economy can be understood exclusively or even primarily in these terms. As noted, agro-TNCs have been the chief agents and beneficiaries in the development of the global food economy with interests, growth and flexibility that now transcend national food economies; even in the world's wealthiest agro-exporting nations many more farmers have been hurt than have benefited from agrarian change and global market integration, and there are still-greater chasms within most farming populations of the developing world. Yet the lens of national and regional disparities is still a useful basis for understanding global productive imbalances, as well as being critical to making sense of why multilateral trade regulation – scripted by governments – so plainly reflects and entrenches corporate interests, as will be seen in Chapter 4. As Canada's former Minister of Agriculture put it, trade agreements are about much more than trade; they're about the rights of corporations 'to do business the way they want, wherever they want' (quoted in McNally 2002: 29).

Chapter 4 examines how disparate agricultural systems were bound together under the authority of the WTO's Agreement on Agriculture (AoA) and why this institution is such a focal point in the battle for the future of farming. While champions of the WTO have couched the issue in vague promises that a rising tide of global trade will 'lift all boats', including (some even suggest *especially*) those of the poorest (in its most forceful conception this argument has sought to frame opposition to trade liberalization as regressive and 'anti-poor'), this chapter explains why in fact the contrary is true of WTO's AoA. Having been largely scripted by rich-country governments, particularly those of the USA and the EU, whose agro-TNCs had an inordinate influence on negotiating priorities, the AoA serves as a veritable supranational constitution for these corporations but works against the interests of the large majority of the world's farmers.

A key aim here is to explain the basic imbalances in the structure of the AoA. These imbalances are commonly framed by the contradiction that trade liberalization was pursued at the same time as the immense agro-subsidies in the USA and the EU were largely sanctified, and subsidies have been a lightning rod for criticism of the AoA and of the WTO more broadly. The nature of these subsidy regimes is indeed part of the unevenness in the global food economy, bearing on the chronic grain-livestock surpluses and long-term depression of global prices for food staples, but this is only part of the picture of what's wrong with the AoA and risks obscuring other ways in which the growing power of agro-TNCs was shielded from democratic control, how the stage was set for further trade liberalization and the basic fact that agricultural decision-making was moving much farther away from the land and from social and ecological concerns. The problems with the WTO's role in agriculture are set against the struggles between those who want to contest its authority and those who want to deepen its liberalizing mandate, which famously came to a head in Seattle (1999) and Cancún (2003). Agriculture was central to the collapse of these ministerials and to the protracted stalemate of the Doha (or so-called 'Development') Round, which has left the WTO mired in a crisis of legitimacy with negotiations temporarily stalling again in 2006. The chapter ends by assessing future prospects in the context of a very problematic status quo and the debate between opponents of the WTO about whether their efforts should be focused on razing or renovating the organization.

Ultimately, when we seek to challenge corporate-led globalization and its institutional fortification, we encounter a highly abstracted narrative in which the competitive pressures of market integration are portrayed as being inevitable and inspiring progress and the base objective centres on – or is reduced to – competing according to comparative advantage. This sort of mystification breeds inertia, the object of hegemonic aspirants, as the reproduction of inequitable systems has always depended not only on rules and force but on having many people convinced that the existing order of things is either right, inevitable or irresistible. Critical political economy aims to shake such inertia. In clarifying the inequities, social costs and ecological irrationality of the global food economy (Chapter 1), understanding how the distorted competitive playing field was constructed and the interests and logic underpinning it (Chapters 2 and 3) and assessing how the regulatory system has been constructed to serve specific interests (Chapter 4), the goal of this book is ultimately to help invigorate alternative imaginations and strategic action, the subject of Chapter 5, premised on the belief that the trajectory of the global food economy is something that is, however powerful, *not* irresistible. This is not a false, contrived or even teleological optimism, as darker outcomes yet are possible in agriculture as in other sectors of the global economy, and in how social inequities and ecological instabilities might be managed by global elites to fortify the status quo. Rather, it rests on the recognition that contradictions are *unsustainable systemic tensions*, and as such are potentially *fault lines* for transformative possibilities. Whether the transformations that emerge out of contradictions and crises are for good or for ill depends upon popular consciousness, social organization and strategic action and initiatives.

It is important that illusions of inevitability are not met with rose-coloured views of either traditional or contemporary small-farm practices, and that concern about the social problems associated with rapid de-peasantization is not held up as a defence of the status quo, as though small farming were some sort of residual, even romantic sponge for poor people. Such defensive reasoning plays to the champions of market integration and industrial agriculture, who are wont to use facile binaries against criticism, such as dynamism and progress versus stasis and backward traditionalism. Recognizing this, the challenge to the corporate control of agriculture gains strength not only in appeals that recognize that small farms have been a historic bedrock of cultures but in understanding how – in an age marked by ecological degradation, hyper-productivity and alienating work – they can be a crucial part of more efficient, ecologically rational agricultural systems that sustain dignified livelihoods into the future, a task that is taken up at the outset of Chapter 5. The chapter and the book then conclude by exploring some of the different scales on which the trajectory of the global food economy must be confronted.

While incredible and increasing concentrations of capital are directing the global food economy and a significant measure of un-consciousness about food prevails in many places, encouraging seeds nevertheless abound and are germinating in the battle for the future of farming. I hope this book can play some small role in helping to nourish them.

ONE

The global food economy: contradictions and crises

Uneven bounty

> The shocking news is that hunger is increasing ... There are now 842 million people suffering from undernourishment in a world that already grows more than enough food to feed the global population ...
>
> It is an outrage that in the 21st century one child under the age of five will die every five seconds from hunger-related diseases ... Hunger will kill more people than all wars fought this year. Yet where is the fight against hunger? ...
>
> Hunger has increased, rather than decreased since 1996. This makes a mockery of the promises made by Governments at the World Food Summits held in 1996 and 2002, as well as the promises contained in the Millennium Development Goals. (*Jean Ziegler, Special Rapporteur of the UN Commission on Human Rights, 2004*)

Per capita agricultural productivity grew steadily over the second half of the twentieth century, and while this growth has slowed, there has never been more food available per person on a global scale than there is today (FAO 2002a). As the Special Rapporteur of the UN Commission on Human Rights put plainly, and as the Food and Agricultural Organization (FAO) of the UN has persistently emphasized, the volume of food produced cannot explain the persistence of hunger and undernourishment. In fact, the UN World Food Programme suggests that the volume of food produced is more than one and a half times what is needed to provide every person on earth with a nutritious diet. And yet while the percentage of the world's population living with severe food shortages has declined in recent decades, absolute numbers have grown. Roughly 800 million of the 842 million who suffer from chronic undernourishment live in developing countries. The FAO (2003: 4) calls this a 'continent of the hungry' that outnumbers Latin America or sub-Saharan Africa. Yet while severe episodic famines occasionally make the news in rich countries, the enduring famine of the 'continent of the hungry' is a near-silent one in rich countries. In addition, more than 2 billion people routinely suffer micro-nutrient deficiencies (ibid.; Pinstrup-Andersen 2000), and it is difficult to enumerate just how many people are in various other states of food insecurity that go unmeasured, such as the routine uncertainty of finding the next meal (Magdoff 2004). It is also difficult to derive statistics on the gendered asymmetries of hunger and food insecurity, though this should not obscure the fact that women and girls in many places are further marginalized within households and cultures. But whatever the challenges of measurement, plainly we live in a world where '"hunger amidst scarcity" has given way to "hunger amidst abundance"' (Araghi 2000: 155).

The widely cited UN and World Bank poverty estimates are that 2.8 billion people live on less than US$2 a day, well over two-fifths of the world's population, and that 1.2 billion people live in 'extreme poverty', defined as less than US$1 a day – and some argue that even these enormous figures are serious underestimates (Sanjay and Pogge 2005; Yates 2004). Virtually all of this population lives in the developing world. While universalizing poverty yardsticks have long been wrought with pejorative cultural assumptions, and used to justify dislocation and social upheaval in the name of Western-guided development and modernization, as subsistence needs are increasingly mediated by the money economy – in other words, non-market access to food, water, land and shelter shrinks – poverty lines defined in dollars have more generalized relevance to an understanding of material deprivation. They also help to give some measure to the scale of global inequality. The annual *Human Development Report* by the UN Development Programme (UNDP) has drawn consistent attention to the fact that the top fifth of humanity controls more than four-fifths of its wealth.

The crises of hunger and poverty are especially acute in rural areas, and regionally in South Asia and sub-Saharan Africa. There is a broadly inverse relationship between the scale of agriculture in an economy and the prevalence of hunger; of the population of the 'extreme poor' and hungry, roughly three-quarters live in rural areas and more than seven in ten depend on agriculture for their survival. Agriculture accounts for 9 per cent of GDP and more than half of all employment for the developing world as a whole, and these averages soar to 30 per cent of GDP and 70 per cent of employment in countries where more than one-third of the population is undernourished. South Asia has the largest total population of chronic undernourished at 303 million, nearly one in four, while sub-Saharan Africa has the largest relative population at 194 million, over one in three (FAO 2003). In sub-Saharan Africa, food insecurity is also deeply entwined with HIV-AIDS, as most of the more than 20 million infected live in abject poverty. While the links are not adequately understood, mounting evidence suggests that poverty, food insecurity and land degradation fuel social instability and violent conflict (Pinstrup-Andersen 2000; Berry 1997).

Opposing the problem of hunger and food insecurity is that of obesity. Globally, the population of obese people now actually out-numbers the population of the undernourished. In 2000, the World Health Organization (WHO) identified obesity as a 'global epidemic' given this scale and the fact that it is the main cause of heart disease, the primary risk factor for diabetes, and is a major contributing factor in some cancers and other diseases. The contradictions of obesity and hunger in a world of aggregate surpluses are reflected clearly in the literal and proverbial belly of the global food economy's beast, the United States. Today, 12 per cent of Americans (roughly 35 million citizens) are considered to be food insecure, 4 per cent 'with hunger' (over 11 million) (Nord et al. 2005), while 65 per cent are considered 'overweight and obese' and 30 per cent 'obese', roughly *double* the levels from 1980, as diet-related diseases are pervasive (Hedley et al. 2004; for a more popularly oriented account, see Crister 2004). The US Surgeon General recently warned that obesity would soon be responsible for killing as many Americans each year as smoking (Economist 2003). These perverse poles of the global food economy, obesity and hunger, reflect the basic reality that while food is elemental to life and health it is conceived as a commodity and not a right – food aid and food banks, which reflect a minimalist conception of food rights, notwithstanding – and the motive force of profit prevails over concerns about equity and nutrition.

The place of agro-TNCs in the US food economy, and a fast globalizing model, is likened by Heffernan (2000: 66) to an 'hourglass which controls the flow of sand from the top to the bottom', reflecting how a small number of massive firms are situated between many pro-ducers and even more consumers, a position that gives them 'a dispro-portionate amount of influence on the quality, quantity, type, location of production, and price of the product at the production stage and throughout the entire food system'. On one side, fast-consolidating agro-input TNCs are increasingly controlling seeds, fertilizers, agro-chemicals and livestock antibiotics and compelling the standardization and industrialization of farming techniques, while a handful of very large-scale manufacturers also dominate farm machinery. On the other side, agro-food TNCs are controlling, refining, combining, distribut-ing and marketing what is being produced on farms in expansive new ways, and systematically detaching food consumption patterns from time, space and cultural traditions with long-distance sourcing and distribution networks, sophisticated processing and packaging systems that reduce perishability, and marketing tactics that forge strong consumer loyalty (Friedmann 2004: 1993), part of a more generalized phenomenon Klein (1999) dubs 'branding'. Globally, the rise and spread of processed and pre-prepared meals have implicitly served to undermine the cultural significance of food preparation and consumption, while the marketing efforts of agro-food TNCs in many parts of the world have explicitly aimed to 'downgrade not only local diets per se but also the symbolic value of traditional foods … as culturally inferior' (George 1990: 148). Also related to branding strategies and the de-spatialization and de-culturation of food is the corporate manipulation of place and culture, with many packaged items given an exotic façade that often bears little or no connection to where the food was actually produced and processed: 'Mexican' corn chips, 'Moroccan' soup, 'Mediterranean' pizza, 'Caribbean' fruit punch, 'Cantonese' spring rolls – the list is long. Where food is not so easily decoupled from culture, agro-food TNCs have employed nuanced strategies to conceal global sourcing patterns, such as using various localized brand names in different places (Fagan 1997). As Western entertainment and media exports expand throughout the world they are another nebulous but significant influence on cultural change, lifestyle aspirations and, at some level, diet. Dietary change also relates to increasing urbanization and the desire or need for convenience and pre-prepared foods in fast-paced, fragmented lifestyles (Sexton 1996).

In short, 'consumers' maps of meaning' are being reshaped (Cook 1994: 236) and diets are converging – taking on a '"food from no-where" character' (McMichael 2004a: 11) as they bear less relation to seasonal rhythms and local productive bases. As Friedmann (1994, 1993) has emphasized, the global food economy is characterized by *distance* and *durability*, something which is highlighted by the useful concept of 'food miles'. Food mile calculations account for the distance that food has travelled from land to mouth, which has steadily increased in industrialized nations (Lang and Heasman 2004; Halweil 2002) to the point where estimates of food miles for the average food item in the USA and Canada typically range between 2,000 and 2,500 kilometres. And further, these estimates do not generally include the distance that the inputs which went into producing the food have travelled, which are especially significant for bulky fertilizers, or the petroleum involved in fertilizer and agro-chemical production.

The FAO (2002a) notes how dietary convergence is especially marked in the rich countries of the Organization for Economic Co-operation and Development (OECD), commenting that change in diets has closely followed income growth and has occurred 'almost irrespectively of geography, history, culture, or religion'. But the increasing detachment of diets from space is occurring almost every-where to some degree, with the populous and booming economies of China and East Asia, South-East Asia and India seen by agro-food TNCs as having the biggest capacity for market growth. This process has clear class dimensions, and is most extreme at both ends of the world's economic spectrum: for the world's poorest, who depend on food durables like flour, cornmeal and rice shipped from temperate breadbasket regions, and in affluent supermarkets, North and South, which possess a dizzying selection of fresh and packaged items sourced from around the world – Reardon et al. (2003) discusses the growth of supermarkets in the developing world. And as the health implications of industrial foods become better understood and organic production commands premium prices, given the cost-accounting system in which food produced by machinery, fossil fuels and chemicals appears so much more cheaply than food produced by labour-intensive organic methods, the world's wealthy consumers are those poised to better access the most fresh, nutritious and chemical-free food baskets (Friedmann 2003).

As the globalization of food brings wealthy consumers greater access to more diverse and healthy foods and the poor more refined food durables, there is one especially conspicuous space of dietary convergence between these classes: the proliferation of junk food – soft drinks, packaged snacks and so-called convenience foods that are full of fats, sweeteners, artificial flavours and colouring. Many of the same small farmers in the developing world who cannot earn a decent livelihood on the land can now find a can of Coke, a tin of Nestlé Milo or a bag of Doritos in their rural shops. Fast-food restaurants, so ubiquitous in the American cityscape (Schlosser 2002), are another clear embodiment of this corporate-led dietary conver-gence and the uneven bounty of the global food economy. Fast-food restaurants are spreading rapidly on a global scale, especially in wealthy urban areas of Asia and Latin America and the Caribbean. The exuberant forecast of one of the world's fast-food giants, Yum! Brands, gives a broader suggestion about the corporate strategizing involved in dietary change:

From Hong Kong to Malaysia, a Customer Mania revolution is

taking hold – driving customer loyalty and differentiating the brands ... And there's one thing for certain – this maniacal focus on the customer is driving global growth – growth in sales, growth in profits and growth in new units ... We also had big wins with new product launches last year ... New promotions, such as the 'Hot & On Time or It's Free,' guarantees in Australia and Korea, and the introduction of the Colonel's famous KFC bucket in China have added to our revenue growth ... in China, our fastest growing and most profitable country outside the US ... [our] business volumes and margins continue to be off the charts, and KFC has been rated the number one brand in the entire country![1]

In short, as food production and consumption become bound increasingly tightly within an integrating and uneven global system, small-farm livelihoods are becoming less viable, traditions surrounding harvest, preparation and mealtimes are severed and agriculture is rapidly losing its place as 'an anchor of societies, states and cultures' as it is transformed into 'a tenuous component of corporate global sourcing strategies' (McMichael 2000a: 23).

The industrial grain-livestock complex

The 12,000-item supermarket noted in the Preface presents a compelling impression of diversity, but apart from the globally sourced fresh produce, the majority of the shelves are packed by what George (1990: 44) aptly calls 'commercial pseudo-variety'. The global food economy is increasingly dominated by a small range of crops and farm animal products, with the basic same core of 'raw materials' reconstituted in myriad ways with a range of standardized additives and fabricated flavours. Schlosser's (2002) description of the secretive corporate laboratories cooking up flavours in test tubes helps illuminate one dimension of this very vividly.

As many as seven thousand plant species have been cultivated or collected for food in human history, but this diversity is shrinking precipitously. There was a drastic decline in both the diversity of crop species planted in agricultural systems and the genetic diversity within species (called 'genetic erosion') during the twentieth century, with these declines as great as 75 to 90 per cent according to FAO estimates. Thirty crops now essentially feed the world, providing 95 per cent of humanity's plant-based calorific and protein intake. The world's top ten crops (rice, wheat, maize, soybeans, sorghum, millet, potatoes, sweet potatoes, sugar cane/beet and bananas) supply over three-quarters of humanity's plant-based calories and dominate the world's cultivated lands, and the 'big three' cereals alone (rice, wheat and maize) account for more than half of all plant-based calories and 85 per cent of the total volume of world grains produced (FAOSTAT; FAO 1997). Though the growth of soybeans (an oilseed) has been spatially concentrated in the USA, Brazil, Argentina and China, its scale and central role in the industrial fusion of grain and livestock sectors is such that it is now part of a 'big four' of global crop production. Soybean cake, the hardened mass after the oil has been pressed, provides a protein-intensive feedstock, and this now comprises almost two-thirds of the world's protein feed given to livestock (Ridgeway 2004). Since the 1970s, soybeans have had the fastest growth in land space of any crop, and it is expected that their per capita production will continue to advance over cereals given their key role as livestock feed (FAO 2002a). Between 1990 and 2005 alone, global soybean production roughly *doubled*.

Crop research in the dominant cereals coupled with rising inputs brought a near-tripling of the world's annual grain harvest between 1950 and 1990, and while some of this growth related to new lands brought under cultivation it was primarily rooted in the rising productivity of cereal monocultures, especially the 'big three'. Other staple crops which have particular importance in poor areas or less favourable climates (e.g. cassava, plantain, beans and yam) have received little scientific attention (FAO 1997). On a global scale, grain yield per hectare expanded by roughly 2.4 times during this period, the amount of global cropland under irrigation expanded by more than 2.6 times, and fertilizer use rose more than tenfold (Brown 1996). The model of industrialized, high-input cereal production has revolutionized farming in the world's temperate regions, namely the United States, Europe, Canada, Australia, New Zealand, the southern cone of South America (Argentina, Uruguay and southern Brazil) and the former Soviet Union, as well as in some pockets of the developing world.

Industrialized agrarian landscapes dominated by a small number of grains and soybeans are inseparable from rising livestock production, and this *'meatification' of diets* – that is, the radical shift of animal products from the periphery to the centre of human food consumption patterns – is another fundamental aspect of dietary convergence and the uneven bounty. While the human population has more than doubled since 1950, meat production grew nearly fivefold by volume (Nierenberg 2005; WorldWatch 2004) – implying a near-doubling of meat consumption in the average diet of every single person on earth amid a soaring human population. On a global level, per capita egg production also doubled in this time, as did per capita fish production, with the volume of the latter roughly half that of the total volume of animal flesh produced. Brown (1996: 53) calls rising total and per capita meat production in the second half of the twentieth century 'one of the world's most predictable trends', and as a result of this 'livestock revolution' animal products now comprise 37 per cent of gross agricultural production according to the FAO (Delgado and Narrod 2002; Delgado et al. 1999). Though these aggregate statistics conceal great disparities within and between nations they clearly underline the magnitude of global dietary change.

Meat consumption and production remain uneven, most intensive in the world's temperate regions, but levels are rising quickly across the developing world. The 20 per cent of the world's population living in the world's richest countries consume about 40 per cent of all meat, or 80 kilograms per person, and this is projected to grow still further to 90 kilograms per person (FAO 2002a). Since the 1960s, however, the relative increase in per capita meat consumption has been much greater in the developing countries of the South than in the North, especially in parts of industrializing Asia and Latin America (Nierenberg 2005, 2003; WorldWatch 2004; Economist 2003; FAO 2002a, b). For the developing world as a whole, over the last three decades of the twentieth century the per capita consumption of meat increased by 150 per cent, from 10 kilograms to 26 kilograms, while the per capita consumption of milk and dairy products rose 60 per cent, from 28 kilograms to 45 kilograms (FAO 2002a).[2] Per capita meat consumption in the developing world is now approaching 30 kilograms per person, and is expected to reach 36 kilograms per person by 2020. Even in the world's poorest region, sub-Saharan Africa, per capita meat consumption is projected to double by 2020. At the clear forefront of these trends is East Asia and the emerging economic colossus of the twenty-first century, China. Per capita meat and dairy consumption in China has more than doubled over the past two decades and is expected to reach the levels of industrialized countries by 2020. In 2005, China consumed more meat than did the world's entire human population in 1961 (Nierenberg 2005). On a global level, the FAO (2002a) projects the per capita consumption of meat to continue rising by another 44 per cent by 2030, while consumption of most other food items levels off (see also Delgado and Narrod 2002; Delgado et al. 1999).

As with crops, livestock production is concentrated on a handful of species, with 88 per cent of all animal flesh by volume in 2005 (265 million tons) coming from livestock's 'big three': pigs (39 per cent), chickens (26 per cent) and cattle (23 per cent), with the production

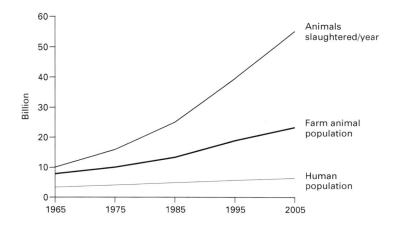

FIGURE 1.1 The meatification of diets worldwide

of pigs and poultry (primarily chickens, but also ducks, geese and turkeys) growing at an incredible rate. Since 1990 alone the global volume of chicken production in the world has roughly doubled, and the FAO (2002a) expects poultry to lead the continuing growth of meat production in the coming decades. Livestock populations understate the growth in production. From 1965 to 2005, the population of pigs on earth at a given time in a year has nearly doubled, reaching 961 million, and the population of poultry more than quadrupled, from 4.2 to 17.8 billion, but because the 'turnover time' of animals has been so shortened by industrial techniques the annual number of animals slaughtered has actually grown at a higher pace. Under industrial techniques, broiler chickens can be brought to slaughter weight in a few months, pigs in as little as six. In 1965, roughly 10 billion farm animals were slaughtered; by 2005 this had risen to more than 55 billion, led by a more than a sevenfold increase in the annual number of chickens slaughtered, from 6.6 to 48.1 billion (see Figure 1.1).[3]

The industrialization of farm animal rearing is the productive basis for the rapid meatification of global diets, with maize and soy the two primary feed crops, often combined. Pioneered in the USA with broiler (meat) and layer (eggs) chickens, Concentrated/Confined Animal Feeding Operations (CAFOs), or factory farms as they are widely known, involve the warehousing of large populations of animals in crowded, industrial conditions, where their growth and biorhythms can be managed and accelerated. Though the export of this model has been relatively recent, on a global scale factory farms are already responsible for 40 per cent of all meat production by volume, a dramatic increase from 30 per cent only a decade earlier. Factory farms are also responsible for a much higher percentage of the global farm animal population than this large and growing volume suggests, since 74 per cent of the world's poultry and 68 per cent of the world's egg production comes from factory farms, there are many more individual chickens than any other farm animal species, and their 'turnover time' is the shortest. The factory farming of pigs is growing at a similar pace, and now accounts for 50 per cent of the world's pig meat (Nierenberg 2005, 2003). Factory farming is even moving offshore, with strong links to industrial farm systems on land. In aquaculture, or fish farming, fish are densely stocked in large pens, fed not only other fish but also grain and farm animal by-products, and given chemical additives and antibiotics to control disease. Roughly a third of the global fish harvest is, in turn, ground up into fish meal and fed to livestock (Nierenberg 2005). With open-water fisheries effectively exhausted and fish farming growing so quickly, the FAO (2002c) projects that the volume of fish produced from aquaculture will overtake the volume caught at sea as early as 2020. A number of fast-industrializing Asian nations are at the helm of the rise in industrial fish farm production. While small-scale, artisanal fish farming (e.g. shrimp, fish, crabs) has a long history in integrated Asian rice paddy systems, the growth in fish exports is not from these methods but from coastal and oceanic pens (Nierenberg 2005).

Having already come to dominate livestock production in much of the industrial North, factory farming has also begun growing quickly in significant parts of Asia (namely China, India, Indonesia, Malaysia, Pakistan, the Philippines, South Korea, Taiwan, Thailand and Vietnam) and Latin America (especially Argentina, Brazil, Chile and Mexico), with China now the world's leading volume producer of animals and home to half the world's pigs. With most of this production oriented towards affluent classes in growing domestic markets, factory farms are typically situated close to urban centres where they compete for water with the sprawling slums discussed in the following section. There is a significant export emphasis to the rise of factory farming in some countries, particularly Brazil, Argentina, and Thailand (Nierenberg 2005, 2003). The FAO (2002a) notes that in recent years industrial livestock production in the developing world 'has grown twice as fast as that from more traditional mixed farming systems and more than six times faster than from grazing systems', and projects this trend to continue into the foreseeable future.

A snapshot of global agro-food trade

Roughly 10 per cent of the world's total agricultural production is now traded across national borders, and as market integration speeds up and thrusts different farm systems into competition with one another, the structural surpluses from the large-scale, highly mechanized (and in the USA and the EU, heavily subsidized) farm sectors of the temperate world have come to dominate global agricultural trade patterns and prices. In 2004, both the USA and the EU produced roughly 17 per cent of the world's agro-exports by value, Canada, Australia and New Zealand together accounted for 15 per cent, and the major South American exporters (Brazil, Argentina, Chile and Uruguay) 13 per cent. This means that 62 per cent of the world's agro-exports in 2004 came from countries that together comprise only 15 per cent of the world's population and only about 4 per cent of the world's agricultural population. This same group of countries accounted for 40 per cent of the world's agro-imports, predominantly the EU and the USA.[4] These industrial surpluses – sourced at prices that bear little relation to the actual economic or environmental costs of production and dominated by a small number of giant agro-food TNCs – effectively establish the global market price for the world's food staples (Rosset 2006). The real prices of the 'big three' cereals, rice, wheat and maize, fell by 60 per cent from 1960 to 2000 (FAO 2002a). The prices of livestock products have also fallen sharply on a global scale, with beef a good example: between 1971 and 1997, the real price of beef declined by two-thirds (Economist 2003). As the FAO (2003: 21) notes, 'depressed world prices create serious problems for poor farmers in developing countries who must compete in global and domestic markets with these low-priced commodities and lack safeguards against import surges'.

By far the largest segment of global agricultural trade is the grain-livestock complex, and these exports are dominated by a small number of countries. The major components of this are cereals, meat, soybeans and dairy products, which collectively comprised 45 per cent of the value of global agricultural trade in 2005, which understates the scale given how these are contained in various processed foods and how cheap and bulky grains comprise a much greater relative volume of agricultural trade than their value represents. The export of cereals relative to total world production has been relatively constant, but this could increase as the FAO (2002a) projects that cereal production in the developing world as a whole will not keep pace with its growing demand, forecasting a 157 per cent rise in the volume of the cereal deficit and a 56 per cent rise in its food deficit as a ratio of consumption by 2030.

The major cereal exports are wheat and maize, with the USA, the EU, Canada, Australia and Argentina combining for almost 70 per cent of all cereal exports in 2004, including virtually all wheat and derivatives (the most traded cereal by value). US grain exports grew fivefold in the second half of the twentieth century and, as will be discussed in Chapter 3, this played a major part in shaping import dependences and consumption patterns in many of the world's poorest countries. The USA is the chief exporter of maize, accounting for about two-thirds of the world's total, followed by Argentina. More than 60 per cent of traded maize is used for animal feed, and maize exporting for feedstock goes hand in hand with soaring soybean exports, which are almost entirely used for feed. Soybean exporting has more than doubled since 1990, dominated almost exclusively by the USA and Brazil. Over 90 per cent of global rice production occurs in Asia, where it is the keystone of food security, and only a small percentage of rice production is traded. Two of the world's largest grain producers are China and India, which together produced almost 30 per cent of the world's cereal output in 2004, but both are minor players in global agricultural trade as they have long sought to manage external trade as part of their food security policies (while pursuing very different agrarian transformations). China has fairly consistently accounted for 5 per cent of the world's total agro-imports and exports, and India only 1 per cent of global agro-imports and exports, though this could soon change rapidly (see Chapter 3). The dramatic growth in per capita global meat consumption has been relatively higher in developing countries, and while this has been met primarily by rising local production it has also led to the rising importation of feed crops as well as meat and dairy products. Pig and poultry are the world's most traded meats, and as with cereals and soybeans global meat exports are highly concentrated: roughly three-quarters of the world total derives from the USA, Brazil, the EU, Canada and Australia, with Brazil's export capacity having grown spectacularly in recent years.

The flipside of the highly mechanized temperate grain-livestock production at the heart of global agricultural trade is the fact that the overwhelming majority of the world's farming population is located in Asia, with three-quarters of the world's farming population, and Africa, with almost one-fifth (see Figure 1.2), where food import dependence is greatest. Most of the world's population growth is expected to be concentrated in Africa and Asia in the coming decades;

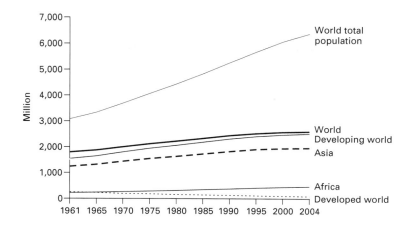

FIGURE 1.2 Agricultural populations

despite the HIV-AIDS pandemic, sub-Saharan Africa will be home to one in three people added to the world population in 2030 and one in two in 2050 (FAO 2002a). Food import dependence is expected to continue growing in most of the world's Least Developed Countries (LDCs), which, as noted, tend to have the greatest relative agrarian populations.

An agricultural trade deficit implies, on a basic level, that a nation's demand for food – not confined to nutrition-based needs, and obviously affected by changing dietary patterns – cannot be fully met by its own domestic production and the agro-export equivalent to its agro-imports. While a deficit may reflect a measure of national food insecurity this is not necessarily so. In fact, neoliberal advocates of global market integration such as the International Monetary Fund (IMF) and the World Bank insist that trade liberalization, the 'discipline' of the market and export competitiveness are the best assurance of national food security, irrespective of the overall agricultural trade balance, and this logic has been repeatedly implanted in their structural adjustment programmes (SAPs). The essential argument of the free market approach to food security is that a nation should concentrate its productive resources on those sectors, agricultural or otherwise, where it has a comparative advantage to maximize the generation of foreign exchange, and open its markets in order to import goods that are produced more cheaply elsewhere (SAPs and the free market approach to food security are discussed in more detail in Chapter 3). But for most developing countries which have narrow, commodity-focused export bases and have endured protracted declines in terms of trade – the 'tropical commodities disaster' (Robbins 2003) – this logic does not hold up well. For countries like Burkina Faso, Bangladesh or Haiti, and the many low-income food deficit countries identified by the FAO which collectively spend roughly half of their foreign exchange on food imports, such prescriptions have widely translated into 'food dependence on world "breadbasket" regions' (McMichael 2000a: 23).

The FAO (2002a) projects that the broad patterns of agricultural trade described here will deepen considerably in the coming decades, with grain and livestock exports from temperate regions entwined with rising food import dependence in much of the developing world. Compared with annual levels in the late 1990s, the grain imports in the LDCs are expected to double by 2020, while meat imports for the developing world as a whole are expected to grow nearly fivefold by 2030, from 1.2 to 5.9 million tonnes (MT), and the annual volume of milk and dairy imports is expected to nearly double, from 20 to 39 MT (ibid.). The competitive pressures and dietary shifts associated with market integration in the highly uneven global food economy ultimately make small farming less viable by reducing earnings per volume output, especially when faced with debts and rising input costs. These patterns and projections are an integral part of an unfolding social revolution in the developing world known as 'de-peasantization', where attention now turns.

The great agrarian question of the twenty-first century

Eric Hobsbawm (1994: 289, 415) argues that 'the death of the peasantry' was 'the most dramatic and far reaching social change' of the twentieth century, cutting 'us off forever from the world of the past', as 'the peasantry, which had formed the majority of the human race throughout recorded history, had been made redundant by agricultural revolution'. Between 1950 and 1990, for the developing world as a whole, agricultural employment declined from 80 to 60 per cent of the workforce (UNHSP 2003), and the world's agricultural population, though still rising slightly in absolute terms, continued a steady decline relative to the world's total population, falling more than 5 per cent from 1990 to 2004. Meanwhile, in the 1990s alone the urban population of the developing world increased by an astonishing 36 per cent (ibid.). Some time in 2007 the world human population of roughly 6.6 billion people will hit a historic milepost: it will become more urban than rural. The UN Human Settlements Programme (ibid.) submits that the world's rural population has likely peaked, and that all future population growth will occur in cities as the world heads towards a population of between 9 and 10 billion by mid-century, where it is generally expected to level out.

While Hobsbawm does effectively draw attention to the scale and speed of social change and urbanization, caution is needed in using terms like 'the death of the peasantry' and 'redundancy'. Small-farm households, after all, still constitute nearly *two-fifths of humanity* (Araghi 2000; Bernstein 2000), and redundancy concedes a measure of inevitability to the current course. Rather than an endpoint, de-peasantization is better understood as a social revolution impelled by unequal distributions of land, the lopsided industrialization and control of agriculture, market integration and the environmental irrationality of increasing food miles. The contention here is that neither this course nor its social fallout is inevitable or stable.

As discussed in the preceding section, global agricultural trade is extremely imbalanced, with a few per cent of the world's farm population responsible for more than three-fifths of the world's agro-exports by value, at the same time as there are many low-income, increasingly food-import-dependent nations with large small-farming populations. Amin (2003) provides a conceptual framework to assess global disparities in per-farmer productivity, identifying three broad classes of farmers in the world. The first are the massive-scale, highly mechanized grain-livestock producers in the temperate world whose total population is in the tens of millions. The second is a more populous but still relatively small group of large-scale farmers who are well positioned within inequitable developing-world landscapes in countries like India, Thailand, Chile and South Africa, and have benefited from industrial methods and inputs. The third group, which makes up the overwhelming proportion of the global agricultural population, are the small-scale farmers who lack access to most new technologies, but might be trapped by rising costs for some inputs like seeds, depend upon human labour, have little or no government support, and often do not have sufficient or good-quality land. Amin (ibid.) estimates that the per-farmer output of the first class is roughly two thousand times that of the third.

Disparities in technology and scale are exacerbated by the agricultural subsidy regimes in the world's richest countries, which together spend over US$200 billion subsidizing their own agricultural sectors, concentrated on the largest farmers (the actual dollar figure attached to rich-country agro-subsidies varies significantly, with the World Bank's oft-cited estimate being US$1 billion *a day*). In contrast, these same countries devote less than US$1 billion *a year* in official development assistance to agricultural development in the developing world (FAO

2003), and between 1988 and 1996 foreign aid for agricultural development to poor countries fell by 57 per cent (Paarlberg 2000) (this is a
point of reference, not to suggest that development assistance has always
been necessarily beneficial, as it has often doubled as an export promotion exercise and entailed contentious assumptions about the path to
modernizing development). As will be seen in Chapter 4, rich-country
agro-subsidy regimes have become a flashpoint for protest over how
global agricultural trade is regulated, though this has been a partial
and potentially distracting target.

With so many small farmers incorporated into market relations
and global market integration deepening in the context of long-term
global price declines and distorted competition, the magnitude of
threatened dislocation is truly staggering. Amin frames this in blunt
terms (2003: 3):

> … agreeing to the general principle of competition for agricultural
> products and foodstuffs, as imposed by the WTO, means accepting
> the elimination of billions of non-competitive producers within the
> short historic time of a few decades. What will become of these
> billions of human beings, the majority of whom are already poor
> among the poor, who feed themselves with great difficulty?

Even a widely dispersed, decades-long industrial boom would not
come close to absorbing so many people into productive employment,
and this is highly unlikely given the patterns of industrial growth
across the developing world (ibid.), to say nothing of the environmental impacts or biophysical constraints of such a prospect. According to
the International Labour Organization (ILO), the informal sector has
dominated recent job creation in Latin America and Africa (Bangasser
2000), and as much as two-fifths of the economically active population
of the developing world are informal workers concentrated in urban
and peri-urban areas, comprising 'the fastest growing, and most
unprecedented, social class on earth' (Davis 2004: 24, 2006). Castells
(1993: 37) describes this as a shift of a significant part of the world's
underclass 'from a structural position of exploitation to a structural
position of irrelevance'. Compounding this is the fact that most developing countries lack significant outlets for external migration, such as
European countries seized through colonialism when capitalism and
industrialization revolutionized their countrysides and enabled the
continent to effectively export its agrarian question (Magdoff 2004;
Amin 2003). Rather, in a great historical twist, immigration policies
in rich countries today are skewed to primarily admit the wealthiest
and most highly educated professionals from developing countries.

So while some ex-peasants might find formal-sector employment in manufacturing or services, the flight to escape rural poverty is pushing many into a new vulnerability in insecure or informal work arrangements in cities; what Araghi (2000: 145) describes as the 'huge urban masses of superfluous people'. Though these 'superfluous' masses might appear to be irrelevant to formal labour markets and systems of accumulation, the scale of global un- and underemployment – coupled with the mobility of capital – acts as a strong global brake on the power of labour and the cost of wages, connecting the process of de-peasantization to broader labour market inequalities. A recent report by the UN's Economic Commission for Latin America (2005) also pointed out the fact that waves of urban migration are often heaviest among the youth, depriving rural areas of their talents and energies and leaving behind an ageing population, an important and underemphasized dimension of this urban migration.

With breakneck urban migration so 'radically decoupled from industrialization, even from development *per se*' in large parts of the developing world (Davis 2004: 9), and racing ahead of growth in adequate employment, housing, infrastructure and service provisioning, the answer to the agrarian question of the twenty-first century appears to be pointing towards what Mike Davis (2006, 2004) evocatively calls a 'planet of slums'. In a detailed survey on the character of the contemporary urban explosion in the South, the UNHSP (2003) estimates that roughly one billion people currently live in slum conditions, which it defines in terms of lacking access to durable housing, safe water, adequate sanitation, secure tenure and sufficient living space. This constitutes almost one in every three urban residents on earth, nearly all of whom are in the developing world. At present, slums comprise 78 per cent of the collective urban population of the LDCs. Regionally, Asia has the greatest absolute slum-dwelling population (60 per cent of the world's total), while sub-Saharan Africa has the greatest relative share of its population in slums (72 per cent) (ibid.). The UNHSP (ibid.) projects that on the current course the global slum population will be 1.4 billion by 2020 and that it will approach 2 billion by 2030, at which point it would represent nearly one in every four people on a planet of more than 8 billion.

The threat of 'slum sprawl' is complex and daunting in the world's two most populous nations, China and India, which are together home to just under two-fifths of the world's population and more than half of the world's agricultural population. Though both have sought to limit agricultural trade, their population and rapid (if uneven) industrialization, economic growth and urbanization have agro-food TNCs eyeing major market expansion potential, and both countries have come under increasing multilateral pressure for liberalization. China's race up the protein ladder towards more meat-intensive diets has turned it into a net grain importer, with increased feedstock import dependence likely in the future. And despite their high rates of industrial growth, tremors of seismic social change abound in both rural China and India, discussed in Chapter 3, which agricultural trade liberalization would undoubtedly magnify.

Finally, it is important to note how this urbanization could lead to a dangerous expediency in food policy. Challenging new food security questions are posed by the rapid expansion of urban poverty and the concurrent decline in people's abilities to access food outside of markets. Thus, while cities have long been seen as poles of development and modernity, in a planet of slums problems of food insecurity and undernourishment are bound to increasingly shift 'from rural to urban areas, even though the prevalence of each of these conditions will continue to be higher in rural areas' (Pinstrup-Andersen 2000: 131). The urbanization of food security problems in slums could, in turn, reinforce short-range national food policies such as trade liberalization designed to ensure access to the cheapest supplies on global markets, since underfed, frustrated and concentrated urban masses tend to be viewed as a more volatile and hence more politically influential constituency than the rural poor (Weis 2004a) by governments fearing that, as Bob Marley once put it, 'a hungry mob is an angry mob'.[5]

Assessing the ecological footprint of industrial agriculture

The destructive and unstable trajectory of the global food economy must be understood not only in human terms but in terms of its impacts on the environment and on other species. Agriculture is a major part of humanity's 'ecological footprint', a conceptual framework that quantifies the land-space needed to meet the resource consumption (source) and waste absorption (sink) demands levied by individuals and societies, and helps to envisage the scale and nature of human economic activity relative to the precarious future of biodiverse life on earth (Rees and Westra 2003; Wackernagel and Rees 1996). To understand agriculture's escalating ecological footprint the discussion starts by reviewing how industrial methods transformed the nature of farming and exact a mounting toxic burden. Agriculture's expanding footprint is then framed within the epochal contexts of climate change and biodiversity loss, which must in turn be understood relative to the political and economic inertia of the present order and the limitations of conventional cost accounting.

Simplification and toxicity Throughout history the long-term viability of farm landscapes has depended upon the maintenance of functional diversity in soils, crop species (and seed germ plasm within species), trees, animals and insects to maintain ecological balance and nutrient cycles. To this end, agro-ecosystems were managed with a variety of different techniques, such as multi-cropping, rotational patterns, green manures (turning undecomposed plant tissue into soils, typically from nitrogen-rich legumes), fallowing land, agro-forestry, careful seed selection and the integration of small animal populations. This conception of farming was transformed by capitalism and industrialization, a transformation which, 'despite many variations in time and place ... shows one clear tendency over the span of modern history: a movement toward the radical simplification of the natural ecological order in the number of species found in an area and the intricacy of their interconnections' (Worster 1993, quoted in Foster 1999: 121). This was made possible by the development and rising use of synthetic fertilizers, agro-chemicals, enhanced seed varieties, farm machinery, concentrated feedstuffs, animal antibiotics and hormones, and the expansion of irrigation systems, which allowed industrial techniques to override previous ecological constraints. Embedded in this has been the extraordinary new dependence upon fossil fuel consumption in the twentieth century, not only in the steadily rising food miles noted earlier and the substitution of combustion engine machinery for animal traction, but with the petroleum demands of proliferating synthetic fertilizers and agro-chemicals. The minimization of on-farm diversity has occurred downwards in scale to the level of soil micro-organisms, detritivores and invertebrates, and in the genetic constitution of plants and animals, and upwards in scale to monocultured landscapes and animal factories where technology can more easily replace human labour.

The simplification of on-farm biodiversity accelerated with rising corporate concentration and control over both farm inputs and outputs. At the input end, packages of seeds, chemicals and fertilizers have been progressively woven together and traditional plant selection and breeding by farmers has moved into laboratories, while pressures on the purchasing end increased demands on the size, appearance and timing of output. Corporate-driven agricultural research treats farmers as recipients (i.e. customers) rather than participants in the process of innovation. Put another way, as farmers become increasingly dependent on manufactured inputs they are losing control over knowledge and with it their experimental impulses and confidence in their own analytical and on-farm problem-solving capacities. Thus, as industrial agriculture expands in scope, knowledge transfer occurs less and less from farmer to farmer or between generations, with older farmers less venerated as a source of applied knowledge, and more from agronomist (sometimes directly employed by agrochemical companies) to farmer. When the culture of farming, to which this generational knowledge transfer is central, breaks down, it is not an easy thing to rebuild. And as the locus of innovation moves off farms and into corporate research laboratories it is destroying the importance of local ecology and knowledge in the process of crop development, one of the most potentially destabilizing aspects of the industrial simplification of agro-ecosystems. Given how global food security hinges on such a small number of crops, the FAO (1997) highlights the importance of conserving the genetic diversity within these crop species and points to the centrality of the world's small farmers in this conservation, at the same time as large monocultures dominate the production volume of all of the world's major crops and the world's seed base is increasingly coming under the control of a handful of very large agro-input TNCs.

The basic promise associated with this transformation has been that 'modern, uniform varieties, under stable, high-input conditions, are both high-yielding and yield-stable' (Wood 1996: 110), a sort of cornucopian narrative – more broadly embodied in the so-called 'technological optimism' of Simon (1981) – in which landscape productivity was seen to be bounded more by human ingenuity than by biological limits and given force by the rising per capita food productivity described earlier. The development of the industrial, high-input model in the temperate breadbaskets and its spread to developing countries in the Green Revolution is credited by some with having accelerated the growth of food production ahead of dramatically expanding human populations in the twentieth century, though condemned by others for its socially polarizing effects, and the belief that technological progress can increase yields still further is used as a justification for research in genetic modification (GM) by its advocates, sometimes spuriously coupled in their arguments with the persistence of hunger and undernourishment (McMichael 2004a).

The most obvious price of an agricultural model built on ecological simplification is chronic toxicity. High-yielding crops grown in industrial monocultures typically require more fertilizer, agro-chemicals and water than do crops grown in rotation or in multi-cropped farms, and the heavy use of inputs serves to both mask problems while creating difficult new ones, setting in motion a treadmill of technological fixes (Altieri 1998, 1995; Gleissmann 1997). Excessive mechanized tillage and the bare ground between planted rows in monoculture fields have led to serious rates of soil erosion, which the FAO identifies as a factor in 40 per cent of all land degradation worldwide, with problems particularly acute in the developing world (FAO 1998).[6] The magnitude of soil erosion is difficult to comprehend given the challenge of thinking on the sort of timescales on which soils are formed (Friedmann 2003), and has long been obscured in industrial systems by synthetic fertilizer. The efficiency with which crops turn nitrogen and phosphorus from fertilizers (along with water from irrigation) into yield gains has, however, been in long-term decline (Buttel 2003; Tilman et al. 2002), and the petroleum that synthetic fertilizer manufacturing depends upon is not an infinite resource.

Monocultures are also more vulnerable to pest infestations, a threat that is typically suppressed by greater chemical usage and which can lead to pest resistance and mutations over time, which in turn tends to be met with more and new chemicals – a cycle that ultimately also affects non-pest species and poses serious risks to human health. In the early 1990s the WHO reported that 3 million people suffer acute pesticide poisoning every year, causing 220,000 deaths (including those caused by pesticide-induced suicides), a figure that does not even account for the impacts of chronic exposure (WHO 1992; WHO and UNEP 1990); more recent WHO estimates now top 250,000 annual deaths. Pimentel (2005) calculates that in the USA alone the annual environmental and societal costs of agricultural pesticides are roughly US$10 billion, with costs measured in terms of their impact on human health, pesticide resistance in pests, crop losses induced by pesticides, bird losses from pesticides and groundwater contamination – see also Pimental and Lehman (1993) for a ranging analysis of the impacts of massive pesticide usage in industrial agriculture. Because plants cannot absorb all the nutrients from fertilizers or agro-chemicals, the excessive nutrient loads and toxic effluence end up running off, leaching and accumulating in groundwater, streams, rivers, lakes and even coastal oceanic waters. The cumulative impact of high-input farming in water is most evident in the hypoxia or 'dead' zone of the Gulf of Mexico, where unnatural levels of nitrogen and phosphorus, stemming mostly from agricultural run-off in the Mississippi river basin, foster the growth of large algal blooms which choke the oxygen from (or *eutrophy*) thousands of square miles of coastal waters and make it uninhabitable for aquatic life (WorldWatch 2004; Buttel 2003).

The industrialization of agriculture and the tremendous productivity gains of the twentieth century have also rested on the transformation of innumerable river and riparian ecosystems through diversionary water schemes, from the micro-scale to colossal canal systems and dams (also prized as a source of hydroelectricity). Large dams have often exacted great environmental and social costs, as they have been a leading cause in the extinction or endangerment of one-fifth of the world's freshwater fish and flooding has pushed somewhere between 30 and 60 million people from their lands (McCully 1996). Irrigation demands have sometimes led to groundwater overuse, typically on a localized scale though in places more extensively. The greatest example of groundwater depletion is in the dry US Midwest, where the industrial grain-livestock complex has been steadily running down the massive Ogallala aquifer, which threatens to raise the possibility of stupendous environmental change as replacement sources are sought (see Chapter 2).

Agriculture is easily the world's largest consumer of water, responsible for 72 per cent of all global freshwater withdrawals, and Pinstrup-Andersen (2000: 133) suggests that 'unless properly managed, fresh water may well emerge as the most important constraint to global food production', an issue that also has great geostrategic importance given the plausibility that access to fresh water may emerge as a key source of conflict in the coming decades. In addition to irrigation supply, the salinization, alkalization and subsidence of heavily irrigated and poorly drained soils can also be a major problem that is often very difficult and, where possible, costly to reverse. The FAO (2006) estimates that as much as 10 per cent of the world's irrigated lands have been severely damaged through waterlogging and salinization as a result of poor drainage and irrigation practices, and also notes that this can contribute to a series of human health impacts through the spread of waterborne diseases, such as diarrhoea, cholera, typhoid and malaria.

The risks associated with ecological simplification threaten to be magnified by GM crops, which differ from millennia of plant selection and the first wave of modern seed enhancement by transferring germ plasm *across* species. This raises the potential for unintended and unforeseeable outcomes such as further pest and weed resistances, viruses and genetic transfers through cross-pollination to non-GM crops and wild plant relatives, as genetically modified organisms (GMOs) can reproduce and recombine once released (Cummings 2005; Kloppenburg 2004; Altieri 1998), and the inevitable drift of seeds makes it near impossible for farmers who reject GM crops in principle to resist the contamination of their fields in practice when such crops are planted in neighbouring fields, an issue that is discussed in Chapter 2. There have been steep adoption curves after 1996 for GM soy and maize in the USA, Argentina and Canada, with some production also growing overtly in China and clandestinely in Brazil (Buttel and Hirata 2003). Though this is a small number of countries it nevertheless means that a large percentage of the world's predominant feed crops are already now grown from GM seed, a transformation that has outpaced both the scientific understanding of long-term risks to ecosystems, other species and human health (e.g. new allergies, impacts on immune systems, threats to antibiotics) and meaningful public discussion and debate in these countries. There have been only a very small number of studies on GMOs and human health, and though troubling concerns have been identified in the countries with widespread GMO diffusion the agro-chemical industry has nevertheless been 'given a blank check by government allowing it to commercialize the technology prematurely, before science could validate the techniques being used or evaluate the safety of the products being developed' (Cummings 2005: 30). Beyond the few major adopters, however, agro-input TNCs have been less successful in keeping the matter from public discussion and democratic control, limiting further GM dispersal and becoming a major fault line in multilateral agricultural trade regulation.

Another major and fast-expanding source of pollution caused by industrial farming practices stems from factory farms and feedlots (Nierenberg 2005; Midkiff 2004; WorldWatch 2004; Ladd and Edward 2002; Marks 2001; Mallin 2000; Silverstein 1999; Tolchin 1998). Incredibly dense livestock populations are major consumers and polluters of water, and these environmental impacts are seen most plainly in the USA, the leader in this fast-globalizing model. In excess of 3,000 litres of water go into producing a single kilogram of US beef, as much as an average American household will use for all of its activities in a month (Durning and Brough 1991), while a factory-farmed pig demands roughly 132 litres of water a day for drinking and the flushing of wastes. A typical US slaughterhouse consumes roughly the daily equivalent of the water demand of 25,000 people, with nearly 1,000 litres of water used per large animal in the slaughter, evisceration, deboning, washing of carcasses and sanitizing of equipment, all of which produces large volumes of waste water full of blood, intestinal contents, fat, grease and cleansing solvents (Midkiff 2004). In 1997, a report for the Senate Committee on Agriculture stated that the annual volume of animal manure produced in the USA (1.4 billion tons) was roughly 130 times more than the volume of human sewage, most of this in vast concentrations on factory farms and feedlots, and noted that animal agriculture is the largest contributor to pollution in 60 per cent of the rivers and streams classified as 'impaired' by the US Environmental Protection Agency (EPA) (ibid.; WorldWatch 2004; Silverstein 1999). As a result of all this, the US EPA describes industrial agriculture as the 'single largest threat to US waters' (quoted in Midkiff 2004: 2).

The waste cycling from factory farms does not come close to approaching how small animal populations on integrated farms were and are used to fertilize soils in rotations with grains, legumes and pastures. Because the volume of faecal matter from factory farms is far greater than what can be sprayed on nearby fields, much gets contained in massive cesspools or 'lagoons', some as great as 7 hectares in size. Up to three-quarters of the nutrient content from factory farm manure does not end up in the crop cycle but gets lost in its storage, treatment and handling (ibid.; Buttel 2003). Containing waste from cattle feedlots is also notoriously difficult. These large volumes of concentrated faecal matter inevitably end up contaminating both the water and air, as lagoons and feedlots and the sprayed manure that does not get absorbed by plants run off into waterways or seep into the ground, another major factor along with the synthetic fertilizer in the excessive nutrient loading which ends up impairing the health of aquatic ecosystems. Some unabsorbed nitrogen also gets released into the air as gaseous ammonia, creating infamous and unhealthy 'smell-scapes' around CAFOs (Nierenberg 2005; Midkiff 2004; Buttel 2003; Silverstein 1999). In addition to the dangers from seepage and airborne emissions, the huge volumes of waste from factory farms are also obviously vulnerable to negligence, accidents and severe weather, particularly the open-air lagoons. One particularly destructive event occurred when a hurricane hit the state of North Carolina, a national leader in factory-farmed pigs, causing 159 million litres of animal waste to enter the state's waterways (three times more volume than the oil spilled by the *Exxon Valdez*, one of the worst environmental disasters in the USA), killing an estimated 10 million fish (Mallin 2000; Silverstein 1999).

Industrial farm animal production poses serious risks to human health in a number of ways. Dangerous diseases are evolving as billions of animals are forced into extremely unnatural lives in cramped and often filthy quarters, and the nature of global poultry production is entwined with the great threat posed by avian influenza, or bird flu. While avian flu has been around for centuries, it has only recently mutated into the exceptionally virulent strain of H5N1 that is now capable of jumping the species barrier into humans, with FAO reports drawing links between its development and the rapid industrialization, growth in scale and geographic clustering of poultry production in China and South-East Asia, as the chicken population doubled in China in fifteen years and trebled in Thailand, Vietnam and Indonesia from the 1980s (Nierenberg 2005). Public health officials worldwide are planning for the eventual mutation of H5N1 into a virus capable of spreading from humans to humans, which would set off a lethal global flu pandemic given the lack of human immunity. The WHO is warning that the occurrence of a large-scale outbreak is a matter of *when* not *if*, and is calling for massive preparation within and between national health systems (M. Davis 2005). Meanwhile, hundreds of millions of chickens have already been culled worldwide, and given that chickens are treated as cheap commodities throughout their life-cycle, with virtually no limit to the pain and suffering that can be inflicted, when they suddenly become an expense to be disposed of rapidly it is chilling but not unexpected to hear that many culls around the world have involved burning or burying alive whole flocks of birds as a means to cut losses (K. Davis 2005).

Another serious public health risk stemming from industrial farm animal production is bovine spongiform encephalopathy (BSE) and the human equivalent, Creutzfeldt-Jakob disease, popularly known as 'mad cow disease' because of the fatal brain wasting it induces. BSE can occur when cattle, a biologically herbivorous species, are given feed containing rendered neural tissue, bonemeal and blood from cattle carcasses, and humans who eat contaminated beef can then be infected. The United Kingdom prohibited 'cannibalistic' feeding practices after its mad cow outbreak, but mad cow scares continue recurring in some places, most notably North America. The spread of factory farming is also linked to a range of infectious food-borne pathogens, including *Escherichia coli* (E. coli), salmonella and listeria (Midkiff 2004; Mattera 2004; Schlosser 2002). As Nierenberg (2005: 49) suggests, however, 'instead of calling for changes in the way animals are raised and meat is processed, many producers and government officials have proposed simply irradiating meat to kill food-borne pathogens and bacteria', which is essentially a means to 'mask filth … from factory style methods'. The localized air pollution from factory farms, particularly the toxic hydrogen sulphide emissions that give pig farms their indelible stench, produces a variety of health problems for workers and nearby residents, including persistent headaches, respiratory ailments, depression, anxiety and fatigue.

Concentrated factory farm effluence has been found to produce the toxic microbe *Pfiesteria pescicida*, which is deadly for fish and can cause a variety of skin ailments for humans (ibid.; Midkiff 2004; Buttel 2003; Silverstein 1999).

The industrialization of farm animal production is inseparable from the meatification of diets, which are typically high in saturated fats and cholesterol and have been identified in epidemiological research as a major contributing factor in chronic health problems such as obesity, cardiovascular disease, strokes, osteoporosis, diabetes and some cancers (IATP 2006; Campbell and Campbell 2005; Barnard 1993; Chen et al. 1990).[7] As well, the long-term health risks posed by the bioaccumulation of rising volumes of the pharmaceuticals used to stimulate growth and suppress disease in concentrated conditions, persistent organic pollutants and agro-chemicals (in feed derived from industrial monocultures, including often GM soy and maize) in factory-farmed animal flesh, egg and dairy products are not well understood and undoubtedly difficult to quantify. Another related human health concern is that antibiotics used to treat humans have been widely employed in factory farming, which can lead to antibiotic resistance in bacteria, fostering the emergence of more dangerous infectious diseases for humans and undermining the effectiveness of some medicines – the reason why the EU banned the sub-therapeutic, growth-enhancing use of such antibiotics (Nierenberg 2005; Buttel 2003).

In sum, there is a rather less savoury underside to the 'pseudo-variety' of the industrial grain-livestock complex:

> … routinely contained in nearly every bite or swallow of non-organic industrial food are antibiotics and other animal drug residues, pathogens, faeces, chemicals, toxic sludge, rendered animal protein, genetically modified organisms, chemical additives, irradiation-derived radiolytic chemical by-products, and a host of other hazardous allergens and toxins. (Lilliston and Cummins 1998, quoted in McMichael 2000a: 31)

Agriculture and epochal ecology While debates persist as to how changes in atmospheric chemistry will affect various climates and weather patterns, sea levels and oceanic currents, ecosystems and bio-feedbacks, the range and resilience of other animal species and agricultural production, there is global scientific consensus that anthropogenic climate change is occurring and a near-certainty that fossil fuel emissions have a fundamental role in destabilizing the earth's atmosphere (IPCC 2007, 2001). The Intergovernmental Panel on Climate Change (IPCC) – the forum established in 1988 by the World Meteorological Organization (WMO) and the United Nations Environment Programme (UNEP) for compiling and assessing the state of scientific research on climate change – has for some time emphasized that the negative fallout from climate change will be unevenly distributed, spatially and in class terms, set to 'fall disproportionately upon developing countries and the poor persons within all countries and thereby exacerbate inequities in health status and access to adequate food, clean water and other resources' (IPCC 2001).[8]

In essence, climate change modelling has projected the threats associated with sea level rises, increased severe weather and changing precipitation patterns to be the greatest in the poorest regions of the world, including many tropical countries, where the resources to respond are the least. Conversely, the per capita greenhouse gas emissions causing climate change are highest in the wealthiest, temperate countries of the world, led by the USA (five times the world average), which is the biggest obstacle to the modest multilateral greenhouse gas emission reduction commitments of the Kyoto Accord. The case for understanding the global inequalities associated with climate change and this inverse burden–responsibility relationship is summarized by Athanasiou and Baer (2002), and the unequal class vulnerability to severe weather hazards was seen unmistakably with Hurricane Katrina in the southern US in 2005, which even the world's richest country managed so poorly. Very recently, however, there has been a marked shift in how the economic fallout of climate change modelling is being understood, with profound economic ramifications for the entire global economy (IPCC 2007; Stern 2006).

The other epochal global context for assessing agriculture's footprint is the rapid loss of biodiversity, which can be understood only on the vast timescale of evolutionary history. While extinctions are a natural process in the course of evolution, for most of the history of life on earth they have occurred in small numbers over very long periods of time (the background rate), save for five 'spasms' in which extinctions occurred in large numbers over short periods of time. Today, as the eminent ecologist E. O. Wilson puts it, 'virtually all students of the extinction process agree that biological diversity is in the midst of its sixth great crisis, this time precipitated entirely by man', through the destruction and fragmentation of natural, self-organizing ecosystems (quoted in Leakey and Lewin 1995: 235; Wilson 2002; Novacek 2001). While there are only estimates about the precise number of species on earth and even larger voids in understanding interrelationships and responses to habitat reduction, modification and fragmentation, which has led to a range of estimates about current and looming extinction rates, this should not detract from the urgency of the matter. A mid-range estimate in the 1990s was that 30,000 species were pushed to extinction every year, or roughly three per hour (Leakey and Lewin 1995).

The range and long-term adaptive capacity of species within drastically simplified and human-dominated landscapes vary greatly, and the outcomes of disrupting ecosystems can often not be fully anticipated nor stability re-engineered. Climate change further magnifies the problem of habitat loss and fragmentation, as rates of change are expected to be much faster than most species could adapt to, and those capable of migrating will find movement difficult over long distances within a human-dominated environment. While plants and invertebrates comprise most of the thousands of species going extinct each year, the global biodiversity crisis has its most profound expression in the endangerment of animals, birds and amphibian species, of which there are much fewer. As natural ecosystems are reduced to smaller and often unconnected patches, other species – especially large animals – can find less refuge from human activity, leaving landscapes of 'ghosts' to borrow Grumbine's (1992) imagery. Macro-scale systems theorizing like Lovelock's (1987) Gaia hypothesis give additional cause for humility about the great unknown that the biodiversity crisis represents, along with both its ecological and moral weight.

But the captains of the global economy are impervious to these ecological and ethical imperatives associated with the extinction spasm and climate change, shielded by an insularity of frightening proportions that is braced by attacks on academic environmental science by corporate-funded think tanks like the Competitive Enterprise Institute (e.g. Bailey 2002) which are routinely echoed in the corporate media. Never has this insularity been put more bluntly than in George Bush Sr's famous remark to the Rio Earth Summit in 1992 that 'the American lifestyle is not up for negotiation', which was echoed a decade later by his son's attempt to explain the US refusal to sign the Kyoto Accord: 'We will not do anything that harms our economy, because first things first are the people who live in America. That's my priority.' In this context it is little wonder that the oil-fuelled productivity, chronic toxicity and increasing food miles discussed earlier, along with the ever expanding footprint of animal agriculture discussed below, have no bearing on the narrow measures of efficiency that are held up by the advocates of corporate-driven industrial agriculture, something that is discussed further in Chapter

5. And because conventional accounting systems fail to account for the atmospheric burden of emissions (the atmosphere being a free sink) or the damage to self-organizing ecosystems and the loss of other species (which go un- or undervalued), the externalization of these mammoth environmental costs must be seen as underwriting how the efficiency of industrial agriculture is reflected in markets.

Modern supermarkets around the world routinely feature food items that have been transported over great distances after having been produced on land that was tilled, seeded and harvested by large fossil-fuel-driven machinery and fertilized and sprayed with petrochemical-derived inputs shipped in from faraway factories, and yet which sell for a cheaper price than could comparable food items produced locally with less input-intensive methods. Whether this reveals something about efficiency and comparative advantage or rather a perverse cost accounting and incentive structure is surely debatable, recalling how the free market approach to food security rests on the assumption that market integration and the 'rationality' of comparative advantage will optimize the efficiency of supply and hence reduce cost per unit for the entire system. Instead, we see this universalizing claim to economic rationality, as held by mainstream market economics more broadly, to be a convenient guise for an ideological framework that allows foundational assumptions to be cast outside the realm of debate or moral concern. Normative economic measures have marginal, if any, capacity to assess the productive efficiency of farming in an ecologically rational way and different frameworks are urgently needed, an argument taken up in Chapter 5.

Additionally, to assume that the costs of industrial foods will continue to be subsidized by the relatively low price of oil is not only an environmental fallacy but could well prove untenable under conventional cost accounting. Most major oil-producing countries have already reached or are fast approaching the halfway point of their oil reserves and the reality of diminishing reserves coupled with the increasing energy needed to extract waning supplies is likely to drive rising costs, a phenomenon that has been called 'peak oil'. When these costs become embedded in transport, machinery use and agro-inputs it will have the most deleterious effect, at least in the short term, on low-income food-deficit countries, which are already spending large portions of their scarce foreign exchange on food imports. Given the extent to which industrial food production is dependent upon fossil energy and petrochemicals, this is a significant component of the great geostrategic tensions embedded in the struggle to control the world's oil supply (Heinberg 2005; Harvey 2003).

To appreciate the widening ecological footprint of animal agriculture requires more elaboration. Since the rise of agriculture humans have simplified and fragmented ecosystems and increasingly affected the lives of animals through farming both directly, in domestication, and indirectly, through habitat loss with the growth of human populations and technologies. Yet while agricultural production inevitably involves some deliberate manipulation and reduction of biodiversity, and though landscapes may have been transformed to the detriment of other species (and eventually often to the long-term detriment of those societies doing the transforming), agricultural systems have never before dominated the lives of individual animals on anything approaching the current scale or degree, nor have they so threatened the existence of other species. The flipside of the biodiversity crisis is that a relatively small number of animal species thrive in human-dominated landscapes at the same time as an even smaller range of animal species are exploited in ever larger numbers and intensity while largely disappearing from visible landscapes into factory farms.

The impoverishment of landscapes thus should be seen not only in extinctions and extirpations, which environmentalists tend to focus on, but in something they typically do not: the rapid growth and confinement of farm animal populations. The first and more obvious reason for this is that this growth has a strong relation to the growing physical scale of human economic activity and resource consumption, part of which has already been discussed and which will be further drawn out now. The second reason is that the ethical dimensions of humanity's footprint do not exist only at the abstract, species level which is commonly held up by environmentalists (i.e. that we should defend the natural areas in order to protect 'wild' species' place on earth), but relate to how sentient life has been commodified; in other words, how the lives of individual animals are dominated to serve human economies and how extreme violence has been systematized and pushed into the unconscious.

As noted earlier, livestock products now account for roughly 37 per cent of global food production, a large, growing and historically unprecedented percentage, if unevenly distributed. But this nevertheless greatly understates the place of livestock production in agricultural landscapes. More than two-thirds of all arable land is devoted to livestock production – roughly one-third of the earth's land surface – either as pasture or to grow feed crops (Nierenberg 2005; de Haan et al. 1997). While some cattle, pigs, sheep and goats graze on land that is not suitable for cultivation or which is left fallow between planting cycles, and poultry birds, pigs, sheep and goats commonly scavenge around small-farm households in developing countries, the soaring global livestock population is increasingly reared in industrial systems, is consuming a rising share of the world's net cereal crop and grazes on extensive areas of agriculturally productive land.

The fact that not all land used for grazing and permanent pasture is suitable as permanent cropland is often used to justify extensive ranching on marginal arable lands, from arid regions to Amazonia, as though this were the only option in a binary equation of pasture or agriculture. In addition to assuming that other species have no claim to habitat on this land, such a case ignores the fact that extensive ranching is a major factor in most areas undergoing desertification (Rifkin 1992). Instead, in light of the global crisis of biodiversity and the scale of desertification, a basic conservation target would be the progressive removal of marginal lands from production and ecological rehabilitation for animal habitat. This is not to be insensitive to local peoples or traditional pastoralists; it is a generalized target and not a dogmatic objective, which must obviously be nuanced in various local settings with justice for local peoples a key goal. Justice often means movement away from marginal areas and redistributing better-quality areas, however, as poor farmers have been widely forced to work low-quality lands by the unequal distribution of the most productive arable land. Small farmers pressed into the frontiers of deserts or the front lines of tropical deforestation are good examples. This assumption in hand, the discussion of efficiency will focus on the efficiency of food production on agriculturally productive land, where the vast majority of animal production occurs.

As a basic rule, 'the resources required to provide a given diet depend chiefly on the amount of animal products it contains' (Gilland 2002: 48). This is because in the process of cycling grains through animals to produce meat, high percentages of plant protein, carbohydrates and fibre are lost. Gilland (ibid.) calculated the ratio of cereal feed to livestock protein product output to be 17; that is, roughly 17 units of cereal feed are used globally to provide 1 unit of livestock product protein output. He derives this calculation from 1999 figures, when 655 million tons of cereals were fed to livestock and 61 million tons of animal protein were consumed by humans, with marine products accounting for 10 million tons of this animal protein and grassland forage fortifying 12 million tons of animal protein, thus resulting in a ratio of 655:39. The US Department of Agriculture has furnished a similar estimate, suggesting that the production of 1 kilogram of beef requires the input of 16 kilograms of grain and soy feed. Precise ratios vary from species to species, with poultry a more efficient converter of feed to edible meat and cattle the least efficient, but in general these nutrient conversion losses in cycling feed through livestock mean the growth of animal agriculture and the meatification of diets require significantly more land to be cultivated per person than would be required for more plant-based diets. Roughly half of all cropland worldwide is used for animal feed, led by large percentages of maize and even larger percentages of soybeans (Nierenberg 2005; WorldWatch 2004; Goodland 1997; Brown 1996; Brown and Kane 1994; Durning and Brough 1991). While rising meat consumption is often treated as a normative part of an improved diet, this is a matter of much contention (Campbell and Campbell 2005; Barnard 1993; Chen et al. 1990).

As well, a sizeable share of the global fish harvest ends up as animal feed in an age when most waters have been harvested to or above their renewable limit. The fast-rising scale of fish farming is requiring additional grain production and levying its own very serious ecological disruptions (FAO 2002c; Ellis 2003). The FAO (2002c) classes 18 per cent of marine fish stocks or species groups as 'overexploited', 10 per cent as 'significantly depleted or recovering from depletion', 47 per cent as 'fully exploited' with 'no reasonable expectations for further expansion' and only 25 per cent as 'underexploited or moderately exploited'. The growth of industrial fish farming is increasingly restricting the access of small-scale fishermen to formerly communal areas and has become a major force in destroying mangroves and estuaries (ibid.).

The basic inefficiency of nutrient cycling means that increasing animal production expands not only agriculture's land-space but its demands on other resources such as water and energy. The additional grain demand and the cultivation of more cropland than would be required by more plant-based diets entail more fertilizer, chemical, water and energy usage on industrial farms. In the USA, for instance, livestock consume roughly 70 per cent of all domestic grains, including an even higher percentage of maize – a crop that alone consumes about one-third of US crop space, 40 per cent of nitrogen fertilizer and more total herbicides and insecticides than any other crop. Resource demands are further magnified by the intensity of water consumption and pollution from factory farms and slaughterhouses and the energy needed to control the temperature of factory farms and run the slaughter process. As a result of the additional grain and the resource budgets of these industrial systems it has been estimated that an edible unit of protein from factory-farmed meat requires 100 times more fresh water and more than eight times the fossil-fuel energy than does an edible unit of protein from grain. In addition, what does not get accounted for in these calculations is the fact that meat and dairy have higher refrigeration demands than most foods, as can be seen clearly in any supermarket. The growth in farm animal populations has another notable impact on the atmospheric 'sink', as the flatulence and waste of farm animals contributes to 16 per cent of the world's emissions of methane, a potent greenhouse gas, and global methane emissions from livestock are projected to rise by another 60 per cent in the coming decades given the continued growth in farm animal populations (Nierenberg 2005; WorldWatch 2004; FAO 2002a; Durning and Brough 1991).

In short, the meatification of human diets and the industrialization of animal agriculture are exacting an ever-growing 'ecological hoofprint' in different ways, and this is increasingly being recognized by environmental organizations and researchers (e.g. Nierenberg 2005; WorldWatch 2004; Silverstein 1999; Lutzenberger and Halloway 1999; Goodland 1997; Brown 1996; Brown and Kane 1994; Durning and Brough 1991). Historically, Western environmentalists have placed great emphasis on containing human population growth, and together with this some are also now stressing the importance of both resource consumption disparities and the important role livestock products play in these. In recent years the WorldWatch Institute has been paying special attention to how rising affluence and meat consumption in Asia are placing increasing pressure on global grain stocks.

But an additionally crucial environmental issue, though it is not often understood as such within mainstream environmental organizations, relates to violence embedded in industrial farming practices and the ethical questions associated with humanity's treatment of, and moral obligations to, non-human animals, as increasing numbers of farm animals are being subjected to lives of immeasurable suffering. Midkiff (2004: x), an author who comes from a pig-farming family in the US Midwest, describes the proliferation of factory farming as an 'ever expanding boundary of suffering and filth'. In factory farms, animals are kept in small individual enclosures on metal grated or concrete floors, allowing excrement to be collected in cesspools and the spaces washed more easily, or in very crowded but shared larger spaces deprived of sunlight and fresh air, and are fed concentrated diets of specialized feed, pharmaceuticals and in some cases hormones. Females are repeatedly inseminated and separated quickly from their young, separation which is a violent act in itself regardless of whether it is recognized that an emotional or biophysical nurturing impulse is being denied. Among mammals this separation is especially swift for dairy cattle and their male calves, which are raised as veal. Routine mutilations such as castration, branding, horn and tail docking and debeaking are done to mask various behavioural problems that emerge from these confined conditions, and are generally performed without anaesthetic. Regan (2004) and Masson (2003) make compelling cases for people to recognize not only the physical pain but also the intellectual and emotional capacities of farm animals, and hence the stress, confusion, fear and sadness they endure in these conditions, arguing that farm animals are the subjects of individual lives that are worthy of moral consideration.

The life of factory-farmed animals culminates in callous transport to slaughter during which they are commonly deprived of food, water and temperature control on long journeys, and face a terrifying final ordeal on the killing floor. At fast-moving slaughterhouses animals are kicked, dragged, thrown or prodded down narrow chutes, watch their fellow beings slaughtered ahead of them amid screams, wailing and the smell of blood, and with the quickening pace of 'disassembly lines' it is not uncommon for animals to be mis-stunned and slaughtered while fully conscious. Factory 'layer' hens arguably face the worst conditions of all farm animals as they are packed in wire-mesh battery cages, in many cases so small they cannot turn around or spread their wings, and their feet commonly grow painfully around the wire as they are immobilized in constant light and cacophony, conditions that have been likened to a 'high-tech torture chamber' (Watts 2005: 531). Male layer chicks, whose genetic make-up makes them uneconomical to raise, are variously suffocated, gassed, drowned, crushed or ground up alive in egg-producing factories upon hatching. Given these conditions and the fact that there are many more individual chickens than any other farm animal species, and that chickens are so heavily factory farmed, Karen Davis (2005) describes this as 'the biggest universe of pain and suffering' humanity has inflicted on another species.[9] Yet while these 'torture chambers' are fast spreading globally they remain shrouded in the social unconscious, hidden from the vast majority of people.

Conclusions

This chapter has sought to bring into focus some of the key issues in what is at stake in the battle for the future of farming. It has done this by exploring the power exerted by agro-food and agro-input TNCs throughout the global food economy, the imbalances in production and trade, the social dislocation that is occurring and projected, and the unstable environmental foundations of modern industrial agriculture at different scales. Agro-input and agro-food TNCs are at the heart of a series of transformative pressures decreasing the viability of small-farm livelihoods: standardizing inputs, production conditions and techniques; replacing labour with technology; shifting control and surplus steadily off the farm; ensnaring farmers in a rising cost–falling price squeeze, reducing their margins; replacing biological diversity on farms with 'pseudo-variety' on supermarket shelves; increasing food miles; propelling dietary convergence (especially for global elites but also, in very different ways, for the world's poor) that is divorcing food from time, space and culture; rearing animals in much greater numbers and confinement; and externalizing untold costs upon the environment, including the expanding 'ecological hoofprint' as more grains are cycled through livestock in absolute and relative terms.

And yet any critical review of the global food economy must also recognize its great bounty. As has been emphasized, never before has so much food been produced on the per capita level, and for most people in industrialized countries and increasing numbers in developing countries, supermarkets appear as a veritable Julian Simon-esque cornucopia, teeming with a far greater range of cheap items than even a few decades ago. This bounty is also coupled with the expansive and intensifying commodification of food and the fact that the social relations, resources, technologies and animal lives behind the commodities are becoming ever less visible and comprehensible to consumers. People in many parts of the world are sitting down to meals with less idea than ever before about where the food in front of them was grown, the conditions under which it was cultivated, reared and processed, what chemicals it contains, who is making the decisions or how it was distributed, much less anything about the broader social and environmental implications of the system through which they get their food. A brand name instead substitutes for most of this knowledge.

Thus, the system is upheld not only by the vested interests such as the corporate executives, shareholders, politicians and large farmers it enriches and empowers, and its ability to provide a surfeit of cheap food, but also by the degree to which its logic becomes either invisible or is accepted as the normative way of doing things, so decreed by the almighty law of competitiveness. And as the logic of food as a commodity produced on large, industrial farms, dominated by huge agglomerations of capital, and purchased in markets becomes more accepted, conceded or invisible, there is no doubt that it acquires an aura of inevitability, akin to Margaret Thatcher's famous claim that 'there is no alternative' to capitalist market integration and the dominance of large corporations within the global economy.

The following chapters attempt to trace the pivotal modern dynamics in how the global food economy evolved and has been institutionally fortified, emphasizing throughout that there was nothing inevitable about this process and that there were and are alternatives.

Notes

1 From a 2002 annual report, accessed in 2003 at <www.yum.com/about/international.htm>. ©2002 Yum! Brands, Inc.

2 Measured between the base years 1964–66 and 1997–99 (FAO 2002a).

3 The statistics given in this chapter are derived from the excellent FAOSTAT database.

4 All world trade totals discussed in this section have been adjusted to exclude intra-EU trade as exports which, when included, greatly exaggerates Europe's place in the global food economy.

5 From the song 'Them Belly Full but We Hungry' (*Natty Dread*, Island Records, 1974).

6 A computer image of the magnitude of the soil degradation problem is produced by the United Nations Environmental Programme, entitled the Global Assessment of Human Induced Soil Degradation (GLASOD). It is available at <www.grid.unep.ch/data/data.php?category=lithosphere>.

7 The Physicians Committee for Responsible Medicine provides valuable sources of public health and nutritional education. See <www.pcrm.org/>.

8 The IPCC's useful website can be found at <www.ipcc.ch/>.

9 For more on Davis's work on behalf of poultry birds, see United Poultry Concerns at <www.upc.org>.

TWO

The temperate grain-livestock complex

Speeding agrarian revolutions

Revolutionary changes in agricultural production once unfolded over the course of millennia (e.g. the domestication of crops and livestock), then over centuries (e.g. the English enclosures and the rise of capitalist agriculture; the ecological impacts of imperialism) and in the twentieth century were compressed into the space of mere decades (e.g. the Green Revolution, factory farming, genetic engineering) (Friedmann 2000). Though the focus in this chapter is on the speeding agrarian revolutions of the twentieth century, the dynamics initiated by the enclosures and the ecological transformations wrought by imperialism are a necessary foundation for understanding the radical simplification, industrialization and integration of the temperate grain-livestock complexes. As Bernstein (2000: 28) suggests, two elemental ways in which capitalism conditions agriculture help to make sense of the structure and ongoing mutations of the global food economy:

> The first is the drive of technical innovation to *simplify* and *standardize* the conditions of agricultural production: to reduce the variations, obstacles and uncertainties presented by natural environments to approximate the ideal of control in industrial production ... The second, and related feature, is the increasing *integration* of farming by capital concentrated upstream and downstream of production on the land.

The rise of agrarian capitalism The changing conception of agrarian property and the enclosure movement in England were at the heart of the historic rupture of feudalism and the origins of capitalist social relations. Feudalism was a deeply stagnant social order, as land was held by title or custom and tribute extracted from the weak by the powerful through obligation and threat of force, stifling social and technological change. For various historically specific reasons, including an especially strong monarchy, a fragmented nobility, weaker customary land rights for peasants and more developed domestic markets, cracks in this order emerged first in England. By the sixteenth century, English peasants were renting land and competing in markets to an unparalleled degree, giving rise to the capitalist motive forces that drive profit-maximizing behaviour: pressure to compete at risk of dispossession, the ability to grow with successful competition, and incentives for productive innovations (Wood 2002, 2000). These incentives for innovation unleashed what were known as the 'improvements' of English 'High Farmers', increasingly scientific efforts at a crop rotation with sheep rearing that expanded wheat yields and marketable surpluses while enhancing the fertility of marginal lands, changes that were, according to Duncan (1996), ecologically sustainable over the long term. Agricultural productivity was also increased by the infusion of new crops such as maize and potatoes from the Americas.

With land becoming a tradable commodity that could be accumulated or lost, where productive landholders could grow in size and less productive landholders would be dispossessed, the stagnant nobility–peasant agrarian class structure of feudalism was giving way to a new agrarian class structure of landlord, tenant farmer and wage labourer. In addition to growth through competition, pressure to grow also led to the expansion in cultivated land. The most basic way to expand cultivated land was to clear forests, which imposed great pressure on wildlife, encapsulated by the celebrated extermination of the last wolf in the late seventeenth century. It is very telling that England, the birthplace of capitalism and the Industrial Revolution, has an environment thoroughly dominated by humans with a few small parks and virtually no self-organizing ecosystems.

The second and more conflictive way to expand cultivated land was to enclose (i.e. privatize) the communally held lands with shared access rights, the commons. The enclosure movement, which had begun centuries earlier, gained steam from the sixteenth century onwards as competitive landlords sought to expand their holdings by expropriating the subsistence rights of others. In short, an agricultural system emerged in which productivity was rising amid the loss and consolidation of smallholdings and a rapid erasure of common property, the most important defence against competition. This *accumulation by dispossession*, so central to the rise of capitalism, created a historic new social class: property-less workers, separated from access to the resources needed to sustain themselves and therefore forced to sell their own labour to farms, factories and mines in exchange for a wage in order to survive (Wood 2002, 2000).

It is significant to remember that the enclosure movement and the increasing concentration of land did not occur without resistance. This dispossession was one spark in the English Civil War (1625–49), and when the monarchy was toppled but private property further sanctified the Levellers and the Diggers emerged to demand more radical social change. The Levellers' struggle for universal suffrage and social and religious equality was suppressed as the war was ending, while the Diggers cultivated their claim to the commons while demanding the universal right to access land and the redistribution of royalist property to the poor, an act of non-violent civil disobedience that was also quickly overwhelmed (Hill 1972). In important respects, these struggles of the Diggers and Levellers at the cusp of agrarian capitalism continue to be reflected in the contemporary battle for the future of farming that is explored in Chapter 5.

By 1700, roughly two-fifths of the English population had been pushed out of farming, a far higher level than anywhere in the world, providing the labour force for England's Industrial Revolution and the soldiers for its imperial army and navy while England's colonies were, in turn, generating immense raw materials stoking its industrialization. For many, including Marx, this push off the land was – however uneven, often ruthless and enmeshed in miserable wage labour conditions – a historically progressive break because it released peasants from the shackles of feudal obligations and unleashed new scientific and technological dynamism. Peasants may have had direct access to land to meet their own needs in feudal societies but with surpluses devoted to enriching landlords and the system held together by force and various forms of coercion, innovation was stifled and peasant households often lived in persistent states of food insecurity and undernourishment, low life expectancy and threats of famine and epidemic disease.

Yet while Marx celebrated the triumph of capitalism over feudal orders and the technological innovation, creative impulses and expanded productivity that were set in motion, he also famously understood the destructive and polarizing dynamics of the new system. Marx explained how, under capitalism, the normative objectives of competition, accumulation and growth relate to a series of relentless tendencies: the incorporation of people into wage relations exploited by owners of capital; the concentration of vast agglomerations of wealth and control over time; the alienation of labour as work is broken down into ever smaller bits of the production process; the relentless conversion of nature into resources; and the imposition of these systemic imperatives upon other societies and environments. The expansionary force of capitalism in transforming ever more labour, land, nature and other animal species into commodities – that is, things whose value is arbitrated by market forces – is beautifully encapsulated by Wallerstein (1997) as the tendency towards 'the commodification of everything'. It is also, as Watts (2005: 528) reminds us, an exceptionally recent process, as 'the full commodity form as a way of organizing social life has little historical depth; that is to say it appeared in the West within the last 200 years'.

The enclosure of the temperate grasslands British and later European capitalism thrived as accumulation by dispossession was projected on a global scale with imperialism drawing more labour, lands and resources into capitalist circuits of accumulation, akin to a great global enclosure. The rise of capitalism in Europe systematically fed its imperial appetite, not only through pressures to acquire new resources and develop new markets, but also because of the immense social tensions welling up from below as the displacement of peasants from the land was outpacing jobs in the factories and mines, even during a period of rapid industrialization, part of what would become known as the agrarian question. In addition to the fates and potential volatility of so many displaced peasants, there was the question of where the end point of the capitalist transformation of agriculture would be and whether capitalism in its purest form of owners of capital and workers would prevail or whether some other class of independent petty commodity producers would persist somewhere in between – because in spite of the agrarian origins of capitalism, the expansive and biological nature of farming means that the ability of capital to monopolize production is weaker in agriculture than any other productive sector (the ability of capital to progressively 'proletarianize' agriculture in spite of these challenges will be examined later in the chapter).

The essential role imperialism played in exporting the tensions associated with the social convulsions in Europe was summarized in blunt terms by the arch-imperialist Cecil Rhodes in 1895:

> I was in the West End of London yesterday and attended a meeting of the unemployed. I listened to the wild speeches, which were just a cry for 'bread', 'bread', and on the way home I pondered over the scene and I became more than ever convinced of the importance of imperialism ... My cherished idea is a solution for the social problem, i.e., in order to save the 40,000,000 inhabitants of the United Kingdom from a bloody civil war, we colonial statesmen must acquire new lands to settle the surplus population, to provide

new markets for the goods produced in the factories and the mines. The Empire, as I have always said, is a bread and butter question. If you want to avoid civil war, you must become imperialists. (quoted in Lenin 1990 [1917]: 79)

Another important dimension of British (and European) imperialism is contained in Rhodes' comment, albeit in a different way than he intended. The phrase 'bread and butter' used here to proverbially denote a fundamental aspect of imperialism could instead have been used literally, as empire also involved a very direct bread-and-butter question: the establishment of wheat (bread) and livestock (butter) were indispensable to Europe's eruption into the world, particularly in the temperate regions of the Americas and Australasia. Where they could be planted, wheat along with feed crops like barley and oats were introduced with livestock. The image of the strong British 'beefeater' projecting and defending the empire mythologized this basic dependence (Rifkin 1992), though the Spanish, French and other Europeans also carried old world crops and livestock to all their colonies. The transplanting of wheat and livestock was also imbued with some religious authority in the 'civilizing mission' of empire, plain in the Lord's Prayer ('give us this day, our daily bread') and throughout the Bible, what Cockburn (1996: 16) calls 'a meat-eater's manifesto'.

Crosby (1972: 66) describes how 'the Europeans immediately set about to transform as much of the New World as possible into the Old World'. At the vanguard of the Europeanization of agricultural landscapes were farm animals for both food (e.g. cattle, pigs, goats, sheep, chicken) and draught (e.g. oxen, horses, donkeys). Pigs were typically the leading food animal in initially establishing European settlement, with cattle the most important over time given their value for meat and dairy as well as for hides and tallow. As Crosby (ibid.: 109) describes it:

> Livestock provided not only much of the muscle with which exploitation of America was undertaken, but was in itself an important end-product of that exploitation, and a factor spurring Europeans to expand the areas being exploited ... The champion European frontiersman of the New World was the cattleman. Again and again, the frontier of European civilization advancing into the interior of the Americas has been that of the cattle industry. This was particularly true in the great grasslands.

This rapid growth of livestock populations encroached upon the indigenous peoples, their farms and gardens and the animals they depended upon (Friedmann 2000; Cronon 1991; Crosby 1986, 1972).

Meanwhile, as valuable crop species like maize, potatoes, beans and squash were carried from the Americas to Europe it enriched European farm systems immeasurably, and crops were exchanged from colony to colony through systems of botanical gardens, enhancing productive possibilities. In their tropical and semi-tropical colonies European nations established a variety of brutal labour regimes for extracting mineral wealth and producing a narrow range of agricultural commodities to serve factories and palates at home, but because Europeans generally fared poorly in these climes systems of rule and wealth extraction had to be maintained with relatively few 'settlers'. So when Europe's agrarian question began to intensify, its export was largely confined to the 'neo-Europes' of the temperate world. The tide of European migration swelled and spread out in the nineteenth century into the temperate grasslands of North America, Australia, southern and east Africa and southern South America. More than 50 million land-hungry Europeans moved to the neo-Europes from 1820 to 1930, a number that 'amounts to approximately one-fifth of the entire population of Europe at the beginning of that period' (Crosby 1986: 5).

In many neo-European landscapes, settlers were encouraged by the large plots of land being made available as governments sought to promote agricultural development and in some instances assert political sovereignty. The moral justification for the enclosures of the indigenous commons rested upon both overt claims to cultural superiority (e.g. race, religion, technological development, etc.) and the fabricated image of the lands as being in a primordial state of 'emptiness', which debased the rights of nomadic and semi-nomadic societies and their understanding and role managing the ecology, and ignored how violence, disease and legal machinations had forced the inhabitants of these lands on to ever smaller reserves. This repeated, resolute blindness was perhaps most stark in the world's fastest enclosure, the Oklahoma land rush, where settlers raced off to claim land at the firing of a gun, a mere half-century after this had been the homeland promised to the already once-displaced Cherokee at the end of their infamous 'Trail of Tears' in 1838.

The manufacturing of steel was a key innovation in the rapid agricultural development of the temperate neo-Europes in the late nineteenth century, as railways linked distant production to urban markets and export ports, steamships moved it faster, and steel ploughs (manufactured in the USA by John Deere by the 1840s) allowed farmers to tear through the thick matted grasses where iron could not. The steel plough released the rich soils underneath, allowing extensive monocultures on increasing scales and spurring major increases in per capita productivity. As the great grasslands were apportioned, fenced and encased in the railway's steel grid, cattle displaced the buffalo and enhanced pasture and monoculture grains displaced native plant species (Friedmann 2003, 2000; Cronon 1991). With this 'welding together' of vast spaces, rail would mark a new age of enclosures, on a different scale to the stone hedges that characterized the original enclosure movement in England.[1] By percentage of pre-contact area and self-organizing ecosystems remaining, grasslands are one of the most thoroughly transformed of all North American biomes.

From the Argentinian pampas to the Australian grasslands, many examples could illustrate the human and ecological violence of the indigenous enclosures that paved the way for the rise of the new grain and livestock complexes in the temperate neo-Europes, but their culmination in the USA is one especially apposite case. As the campaign waged by the US military against the indigenous peoples of the Great Plains (euphemistically referred to as the Plains Indian Wars) was winding down in the late nineteenth century, a spiritual leader named Wovoka emerged. Through a ghost dance Wovoka connected the subjugation of his people with the devastation of the landscape and of the buffalo, an animal that was at the heart of Sioux culture and of the plains ecosystem and which had been virtually exterminated for sport, its hides and to make way for the settlers' cattle. The ghost dance sought to bring hope in the mutual rebirth of the Sioux and of the open grasslands ranged by great herds of buffalo, and as it grew into a small movement the US army sought to crush the subversive narrative it contained by interning a group of Minneconjou Sioux at Wounded Knee (South Dakota) in 1890. There, events degenerated into a slaughter of defenceless Sioux men, women and children, a massacre that is commonly used to mark the final 'closing of the west' in the USA (Brown 1970).

Another definitive symbol of the transformation of North America's open grasslands into an industrial grain and livestock landscape was the opening of the Union Stockyards in Chicago in 1865. At the Union Stockyards, cattle driven and carried by rail from the west were killed and processed in a massive abattoir and meat packing complex which 'perfected the production line slaughter of living creatures, for the first time in the history of the world' (Cockburn 1996: 26; Cronon 1991). The brutal working conditions and cruelty towards animals were lucidly portrayed in Upton Sinclair's famous 'muckraking' novel *The Jungle* in 1906 (Sinclair 1981 [1906]) and the ensuing public response – which focused upon improving labour and food safety standards – foretold how easily the violence towards farm animals would slide into the social unconscious as it escalated throughout the coming century.

Soil mining, through-flow industrial agriculture and the treadmill of technological fixes Large-scale grain and livestock production had tremendous synergy with manufacturing in the USA, from industrial slaughterhouses to John Deere's expanding array of farm machinery and implements, as well as with the cheap and abundant food being supplied to burgeoning cities. Yet while early confidence in the bounty of extensive monocultures in the USA and other neo-Europes was bolstered by the per capita productivity gains in the neo-European temperate breadbaskets in the late nineteenth and early twentieth centuries, this bounty was nevertheless built upon a precarious foundation. Prairie ecosystems had evolved to the dry but high-intensity rainfall climatic conditions, with thick grasses to protect soils and retain moisture, and with no historical precedent on the landscape farmers did not comprehend the erosional vulnerability of monoculture grains, which leave so much soil exposed between planted rows. Instead, extensive farming practices were shaped by competitive pressures and conceptualized within a very short time horizon that increased the immediate costs of investing in more labour-intensive soil conservation practices.

In other words, farmers were being systematically compelled to maximize yield and compete in the short term without due investment in the material basis of their productivity, soil fertility, meaning that they were effectively mining the soil over the long term as soil nutrients were used faster than they were replenished (Friedmann 2000). Berry (1995: 1–2) frames this cost accounting system as 'drawing, without sufficient repayment, against an account of natural fertility accumulated over thousands of years beneath the native forest trees and prairie grasses'. And as the soil-mining surplus grain production from the temperate neo-Europes increasingly entered European markets from the 1870s onwards it dampened food prices for Europe's growing urban and industrial working-class populations, providing a further boost for industrialization there, as well as heightening competition for European farmers. This, in turn, also shortened time horizons and raised the perceived costs of laborious soil conservation methods, which was particularly destabilizing to England's advanced

High Farming rotational system (Friedmann 2000; McMichael 2000b; Duncan 1996). But the illusions of boundless fertility came to a crashing halt with the Dust Bowl in the 1930s as large areas of the dry US west were pushed to near ecological bankruptcy.

The basic science of soil mining had already been identified in 1840 by Justis von Liebig, a German chemist who chemically analysed plant tissues to determine the nutritional needs of plants and explained how specific biochemical deficiencies (especially nitrogen) limit production (Foster and Magdoff 2000). In *Capital*, Marx drew upon the work of Liebig in theorizing 'progress in capitalist agriculture' as being 'the art, not only of robbing the labourer, but of robbing the soil' and ultimately undermining its productive stability (quoted in ibid.: 49). For Marx, the Dust Bowl would surely not have come as a surprise. Liebig, however, had also set off the search for a technological fix for soil mining that would extend the stability of capitalist agriculture indefinitely. While collected experimental wisdom had led farmers in many parts of the world to employ various nitrogen-enhancing and other nutrient-cycling soil management practices (e.g. multi-cropping patterns, green manures, rotations and fallowing and the pasturing of small livestock populations on fallow land and crop stubble), understanding the biochemistry of declining soil fertility of some farming practices such as extensive monocultures was a crucial discovery, for it allowed scientific research to develop inputs (especially nitrogen and phosphorous fertilizers) that could override ecological limits and accommodate capital-driven practices instead of new methods accommodating ecology as in English High Farming (Duncan 1996). Some sources of nitrogen fertilizer occurred naturally. Liebig's discovery turned the accumulated nitrogen-rich seabird ex-crement (guano) on islands off the coast of Peru into veritable gold, which was exported to Europe and exhausted within a few decades. Peru's fertilizer boom and bust were followed by a similar process in Chile, and Chilean nitrate production also helped to partially mask the ecological instability of extensive agriculture in North America and Europe until the development of synthetic fertilizers.

In 1908, the German chemist Fritz Haber developed a method of ammonia synthesis by combining atmospheric nitrogen and hydrogen, subsequently commercialized with Carl Bosch and German chemical company BASF, which would come to provide the primary source of nitrogen inputs used in industrial agriculture. Smil (2001) describes the Haber–Bosch process for manufacturing synthetic fertilizer as one of the most influential innovations of the twentieth century for its essential role in the tremendous productivity gains of industrial agriculture. Yet while signs of deteriorating soil health were evident in the declining maize yields in the 1920s and early 1930s in the USA (Berlan 1991), it took the Dust Bowl for the commercial usage of synthetic fertilizers to really erupt. As Friedmann (2000: 493) explains it:

> … prairie wheat farms recovered from the Dust Bowl by deepening the industrial transformation of wheat farming. After World War Two, farmers turned to the markets to replace the lost riches of dark earth … They bought industrial fertilizers, made from fossils stored in the earth over millions of years, and tractors, run by fossil fuels, to replace each season the nutrients that had once seemed inexhaustible.

In the USA, the volume of synthetic nitrogen fertilizer per ton of crop grew more than fourfold in the 1950s and 1960s (Foster 1999). On a global scale today synthetic nitrogen fertilizer accounts for roughly 60 per cent of total fertilizer consumption, more than half of which is applied to cereal crops (Gilland 2002).

While the rising productivity discussed in the previous chapter is a clear indication that the synthetic fertilizer revolution was able to overcome the loss of nutrients from over-ploughing and the lack of ground cover in monocultures, this was a perpetual and not a one-time fix with industrial farms set on a treadmill of fertilizer dependence. Once on this treadmill, repeated ploughing further reduces the diversity of soil micro-organisms, detritivores and invertebrates in terms of species number and biomass, while soil fertility is narrowly defined in terms of the nutrient requirements of specific crops and targeted accordingly (Swift and Anderson 1993; Paoletti et al. 1992). In both its manufacture, which depends on natural gas, and its inseparability from the introduction of the internal combustion engine to agriculture, the synthetic fertilizer revolution was a major factor in binding the productivity of industrial farms to the increasing use of fossil fuels and in deepening the synergy between agriculture and manufacturing. By 1900 fossil-fuelled tractors were already common in the US west, and by the 1920s harvester-threshers were becoming widespread.

These transformations simultaneously expanded the physical scale on which individual farmers could operate, intensified the homogenizing tendencies of scale and devalorized labour-intensive farming methods, as tractors and harvester-threshers could obviously move much faster than draught animals and since large machinery was inimical to multi-cropping. Industrialization had a masculinizing effect on agriculture as the major contributions of farm women and daughters in temperate-farm households were steadily replaced by machines and inputs. It also allowed for the disarticulation of farm animals and crops; with animals no longer needed for draught, the pasturing of animals and the fallowing of land declined and farm animals were progressively simplified into pure food commodities increasingly fed from the grains produced by large monocultures. This latter revolution in livestock rearing was also enhanced by the introduction and diffusion of a new crop, soybeans (originally from China), a legume that had the benefits of fixing nitrogen in the soil, being easily mechanized and providing protein-dense feed for livestock when crushed into meal. Soy had been a relatively localized crop at the beginning of the twentieth century, rare in the temperate grain-livestock complex into the 1920s, but it quickly proved to work so well in industrial rotations with other cereal grains and in feed mixtures that by the end of the Second World War it was fast growing into its present position as one of the world's dominant feed crops in regular rotations with maize (Berlan 1991).

Another of the great technological fixes that allowed capital-driven practices to override ecology in expansive industrial farming was the development of petrochemical-based pesticides, herbicides and fungicides. Historically, large monocultures were a much greater ecological liability, as they were more susceptible to the onset of specific pest populations and rapid infestation, and to pests, weeds and plant fungus establishing permanent niches. As a result, farmers had to manage functional diversity to mitigate these threats, using multi-cropping patterns and frequent rotations to reduce the target food supply of pests and incorporating beneficial species like pest predators, while weeding was unavoidably labour intensive (though what is defined as a weed in a monoculture would sometimes serve a beneficial role in a multi-cropped farm). The rise of agro-chemicals revolutionized the control of insects, weeds and fungi, replacing the need for on-farm diversity and labour-intensive ecological management with a new normative objective: biological standardization. But because repeated, prolonged use of agro-chemicals can bring about vicious cycles of pest and weed resistance and the elimination of natural predators and controls, another input treadmill emerged with agro-chemicals. In an age when large corporations were promising 'better living through chemistry', the application of agro-chemicals began soaring in the 1950s and 1960s, with governments in the West not only regulating but subsidizing their proliferation ahead of any scientific understanding of the ecological impacts. When environmental scientists like Rachel Carson (1962) did emerge, they were met by aggressive corporate efforts to slander and suppress findings, and though the work of Carson and other scientists ultimately helped achieve significant regulatory protections, state subsidization of agrochemicals persists in various forms, including direct cost support, tax exemptions and permission to continue externalizing environmental and human health costs within 'a system of tolerated levels of toxicity on foods' (Dundon 2003: 12; Buttel 2003).

Synthetic fertilizers, new farm machinery and agro-chemicals together shattered the age-old conceptions of agro-ecosystems as closed-loop cycles of nutrients and energy and managed diversity and allowed industrial capitalist agriculture to technologically breach localized ecological bounds and self-renewing organizing imperatives. Instead, industrial capitalist agro-ecosystems were transformed into an open or *through-flow* system dependent upon the continued purchase and application of various external inputs in order to produce commodities for sale (Berlan 1991). In capitalist cost accounting every step along this through-flow system was measured as national economic growth – fertilizers, machinery and implements, agro-chemicals and the fossil fuels on which they all depended – and each step fed the increasing control of corporations over farming. Meanwhile, the loss of non-market ecological cycles was systematically ignored.

Another key piece in the revolutionary transformation of agriculture into a through-flow industrial model, and in some ways a linchpin securing it against reversal, was the most elemental input, the seed; which is why Shiva (1993) likens the commodification of the seed to a final enclosure of the intellectual commons. From the mid-twentieth century onwards, the development of enhanced seeds brought major yield gains which rested upon the heavy use of fertilizer, chemicals and irrigation (the application of this general package of enhanced seeds for high-input monocultures to the developing world was known as the 'Green Revolution' and is discussed in the following chapter). Modern seed enhancement at first resembled the warp-speeding up of millennia of seed selection to achieve the traditional objectives pursued by farmers, namely improvements in yield, nutrient content, hardiness, appearance and flavour, with yield the primary focus. But as large agro-input TNCs came to dominate the process of seed development and patenting, innovations including changing the genetic foundation of seeds were increasingly pursued with the aim of more completely commodifying the seed – which, by its biological nature, is an elusive thing to commodify – and of using new seed varieties to valorize, or induce dependence upon other inputs (Kloppenburg 2004; Shiva 1993).

Enhanced seeds and fertilizer- and chemical-intensive agriculture also depended upon increased water supplies, with massive-scale dams, aqueducts and deep underground water drilling expanding the scale on which irrigation was possible. Again, the USA provides the supreme embodiment of the engineering feats, technological dependence and ecological vulnerability of this transformation. In addition to the rising application of fertilizers, another response to the Dust Bowl was to tap the large stores of underwater aqueducts in the US Midwest, with intensive pumping increasing after the Second World War. The most important source has been the great, ancient Ogallala aquifer, with pumping volume so much greater than its slow recharge rates that depletion is akin to 'water mining'. Reisner (1993) details the vast irrigation systems of the US west, with his striking image of a 'Cadillac Desert' accentuating the dire prospect of the industrial grain-livestock heartland without the current volume of irrigation water. The great bounty of California, the leading agricultural state in the USA, rests upon arguably the world's most comprehensive irrigation infrastructure which has come at an ecological cost of damming and diverting many rivers, most famously the Colorado, and destroying most of the state's once vast wetlands. Future water shortages in the corn and wheat belt of the US Midwest have raised the possibility of massive-scale diversions from the world's largest freshwater lake system, the Great Lakes, which would entail an incalculable ecological transformation (Barlow and Clarke 2002).

The apex of industrial agriculture: factory farming[2] The disconnection and simplification of natural processes in agrarian capitalism and the progressive 'commodification of everything' reach their apex in agriculture in the rise of Concentrated/Confined Animal Feeding Operations (CAFOs) or factory farming. The commodification of some animals has long and complex historical roots that predate and reach beyond the rise of capitalism, as millennia of domestication drew a small range of animal species within humanity's sphere of moral concern and moved others farther from it to varying and culturally relative degrees. To take a basic example, dogs are revered pets in some cultures but eaten in others, while two of the world's most consumed animals, pigs and cattle, are sacred in major religions. But for the purpose of this discussion, it is enough to note that throughout most of history even those animals reared for their flesh, milk, eggs, hides and wool were raised in relatively small numbers and would have been individually known to the farmers and pastoralists raising them. Whether or not there was an emotional bond in this relationship would obviously affect the treatment to a considerable extent, but in a general sense farm animals' lives were more directly connected to the land and had value for a longer period for various reasons that include: their role scavenging, grazing and cycling nutrients through manure and providing a key source of nitrogen; their muscle power; slower reproductive and productive (e.g. milk and eggs) rates in nature; and the limits of technology. As noted, however, in industrialized agriculture farm animals have approached the status of pure commodities cut off entirely from agro-ecosystems, transformed into economic objects for human use without moral standing or rights, housed in previously unfathomable densities with their most elemental behavioural needs subjected to pressures of cost minimization and their treatment restrained by the weakest of legal protections. In some countries (including the USA, the world's leading producer and exporter) poultry birds are legally defined as property and do not have any enforceable legal protections, while the minimal legal protections that other farm animals have on paper are often very far from routine practice and enforcement.

Factory farming first began when the relatively small-scale warehousing of chickens in the eastern USA quickly proved to be profitable in the 1920s. In the 1930s factory farm pioneer John Tyson was developing the vertical integration of feed, hatchery and chick distribution, outsource growing, slaughter and processing that now widely characterizes industrial poultry production, and by the 1950s and 1960s the company ownership of 'inputs' (feed and chicks) and 'outputs' (eggs, meat) and the contracting out of the growing process to 'farmers' began to intensify. As with seeds, species of chickens were subject to increasing genetic manipulations to enhance productivity, creating animals defined by their productive traits – layers and broilers (genetic engineering has since pushed the commodification of sentient life to ever further extremes). Layers were put into 'battery cages', small wire enclosures where they can't even spread their wings or turn around, and forced to live in constant light to speed up laying rhythms, while broilers were densely packed into long warehouses. The scale of factory farms grew to the point where an *average* building now contains ten thousand birds.

The problems associated with deviant behaviours, by-product disposal and disease threats were not addressed by modifying the crowding of factory farms, but rather ecological limits were again overridden by technologies in the form of assembly-line mutilation, pharmaceuticals and large fans. In order that chickens would not hurt each other (i.e. damage property) with their pecking as they struggled in unnatural conditions, a system was devised to rapidly de-beak day-old chicks, which is performed rapidly without anaesthetic, and because beaks are full of nerve endings it causes short-term trauma and can leave birds in chronic pain. Since layer chickens were genetically manipulated for egg production and not efficient weight gain, males became an uneconomical by-product to be disposed of in the most cost-effective way possible, which has meant they are commonly ground up alive and made into fertilizer or discarded in large piles and left to die. Faecal waste collected in cesspools and lagoons has also been used in fertilizer, though, as discussed in Chapter 1, the volume of waste produced from a factory farm exceeds what proximate fields can handle. To address the acute vulnerability of chickens to respiratory and other diseases when housed in such densities and surrounded by and often in contact with so much faecal matter, as well as the tremendous contagiousness of these conditions, the system became dependent upon sub-therapeutic levels of antibiotics in feed, automated chemical disinfectant sprays and industrial venting systems to regularly flush out contaminated and reeking air. Some drugs proved to have the profitable side effect of speeding weight gain, and growth hormones were also designed for feeds (the EU has banned the use of growth-promoting antibiotic use in animal feeds because of its effect on the effectiveness of human medicines, and in general has stricter standards on hormones and antibiotic usage than the USA). As a result of these various innovations the 'turnover time' of chickens has shortened dramatically; contract farmers can bring broiler chickens to slaughter weight (2 kilograms) in six to seven weeks, while layer hens bear their torturous fate for longer, and are generally 'spent' (i.e. their egg production slows down) and killed somewhere between one and two years of age. A standardized slaughter weight is important to the rapid speed of the kill floor and disassembly lines.

After chickens and other poultry birds, the animal with the biophysical characteristics most conducive to intensive confinement is the pig, which began to be moved into factory farms in the 1960s with many techniques borrowed and modified from chicken rearing. In factory farms, breeding sows spend their entire lives impregnated or nursing their piglets in tiny gestation crates (less than 2 square metres per hog) on floors of grated metal or concrete in which they cannot turn about. Piglets can be weaned in as little as three to four weeks, at which point sows are re-inseminated and the piglets sent off to finishing barns, where they are fattened to a standardized slaughter weight with unprecedented speed, as genetic manipulations and concentrated feeds laden with antibiotics and hormones

The grain-livestock complex

have made incredibly rapid weight gain possible. As with chickens, systematic mutilations (tail docking) and sub-therapeutic levels of antibiotics are used to override deviant behaviours and disease threats resulting from confinement, while the enormous waste lagoons are an economical solution to by-products only since their colossal pollution burden continues to go unaccounted. Cockburn (1996: 39) describes the nature of industrial pig farming in North Carolina, one of the leading pork-producing states in the USA:

> Its reeking lagoons surround darkened warehouses of animals trapped in metal crates barely larger than their bodies, tails chopped off, pumped with corn, soy beans and chemicals until, in six months, they weigh about 240 pounds, at which point they are shipped off to abattoirs to be killed.

Cows are still partially reared on pastures though the production of both dairy and beef cattle has also been increasingly industrialized and concentrated. Dairy cows are kept constantly impregnated in order that they can be maintained in an unnaturally stable state of nursing (i.e. milk production), with the production per cow rising significantly over the course of the twentieth century through enhanced insemination techniques, breed development, antibiotics and, in some places, hormones such as the controversial recombinant bovine growth hormone (rBGH) used in the USA which induces greater milk production. Upon birth dairy cattle have their female calves weaned quickly and their male calves instantly removed and sentenced to solitary crates so small they can barely move so as to inhibit muscle development before they are killed for veal after three to four months. Beef cattle are still widely grazed, though many are now also commonly fattened on densely stocked feedlots. There they are given concentrated feed to approach slaughter weight more quickly, which in some cases is laden with growth hormones and other industrial by-products, sometimes including, as noted in Chapter 1, the rendered parts of other cows.

In short, the industrialization and soaring scale of farm animal production in the second half of the twentieth century represent one of the most profound of all the sweeping and speeding agrarian revolutions that have given shape to the uneven global food economy. These industrial innovations have created factories that bear no connection to the landscape except as sources of contamination, pivot on the rising grain yields discussed earlier and, as will soon be seen, have gone hand in hand with rising corporate control and the incredible polarization of agricultural production.

Managing a state of chronic surplus[3]

In response to the Dust Bowl and the Great Depression the Roosevelt administration established the Agricultural Adjustment Act, a major agriculture support programme in the 1930s, as part of the New Deal. The proclaimed goal of the programme was to lift up and secure the livelihoods of family farmers, the health of the land and the future of rural communities. Those who see the intent of the programme largely in this light tend to draw attention to the price supports, the subsidies given for production controls and the conservation initiatives, such as payments for removing fragile and less productive lands from cultivation and for establishing better soil management practices. The basic logic of production control stems from the fact that even a modest overshoot of supply over demand tends to place great downward pressure on farm-gate prices and hurt farmers, and this was a significant concern before the Dust Bowl and rose considerably with the recovery and boom afterwards, at the heart of what Cochrane (2003) calls the 'curse of American agricultural abundance'. The hope with supply restraint is that successful programmes can secure and stabilize reasonable prices for farmers and enhance their ability to plan their production into the future.

The farm programmes of the New Deal and those which followed did have an important role in revitalizing agricultural production in the USA along with the rising use of fertilizers, agro-chemicals, machinery, irrigation water and scientific yield enhancements. But they did not strengthen family farming or vibrant rural communities as they ostensibly undertook to do, and supply management programmes did little to abate fast-rising agricultural surpluses in grains and oilseeds. In fact, to the contrary, the subsidy regime rooted in this period greatly exacerbated the surpluses and the polarizing trajectory of American farm production. Berlan (1991) insists that this was no accident and urges against seeing the New Deal farm programmes as a well-intentioned, family-farm-centred response to managing American agricultural abundance gone bad. Rather, with many of the key architects themselves prominent figures in large agribusiness companies, he argues that the programmes were designed much in the spirit of their ultimate effect; that is, to solve the overproduction problem by establishing 'a totally new food system based on the transformation of grain into meat' and ultimately to 'foster capital accumulation in the emerging agribusiness complex increasingly dominated by large corporations and banks' (ibid.: 116, 127–8). This emerging agribusiness complex had no interest in supply management that would keep prices higher for farmers, but instead sought to have grain production locked into a chronic, expanding state of surplus in order to cheaply source raw materials (and hence increase retailing margins) for refined cereal products and feedstock in expanding value-added livestock operations, and enhance export competitiveness with which to forge new markets.

On the demand side, consumption of livestock products was rooted in a number of factors. The post-war economic boom and Keynesian distributive measures had brought rapid and across-class growth in disposable incomes, and the increasing consumption of meat and dairy was naturalized as an object of this rising affluence. Corporate lobbyists from the meat, egg and dairy industries essentially designed the 'four food groups' (meat, fish, eggs, beans and nuts; dairy; grains; fruits and vegetables) for government agencies in the 1950s. With this nutritional guideline embedded in public health education, Americans were taught from childhood that the heavy consumption of livestock products was central to a good diet. With slight modification, the basic government nutritional advice emphasizing diets high in animal-based protein has largely followed these lobbies against growing evidence to the contrary and the rise of a national obesity epidemic (Campbell and Campbell 2005; Nestle 2002). Very successful corporate marketing efforts also tapped into the societal changes and fast-paced lifestyles and gave shape to this dietary revolution, with the development of TV dinners setting off a widening array of packaged, ready-to-eat meals and the launch of fast-food restaurants like McDonald's, Kentucky Fried Chicken and Carl's Jr. initiating what would become such a central part of the US palate, culture and urban and suburban landscapes, a rise chronicled so well by Schlosser (2002).

By the end of the twentieth century Americans were consuming less grain products per person than they did a century before but significantly more meat and dairy products (WorldWatch 2004; Bente and Gerrior 2002), and nearly half of the total revenues from agricultural products in the USA came from livestock and livestock products (Gilland 2002). The rising per capita consumption of meat was led by poultry, with the 1930s level of per capita poultry consumption (6.8 kilograms) rising nearly fivefold in half a century (32.9 kilograms by the mid-1980s), while the rising per capita consumption of dairy products was led by cheese, which grew eightfold in the twentieth century (Bente and Gerrior 2002). In the USA today, more than 70 per cent of all grains produced are fed to livestock, and the US agricultural census notes that fourteen times more land is given to the production of hay, a pure feed crop, than is used in growing vegetables consumed by people (WorldWatch 2004). For the champions of industrial agriculture, the fact that so much grain can be cycled through growing livestock populations to provide historically unprecedented levels of per capita meat consumption is a mark of progress, 'improved' diets and the robustness of the model.

Despite the rise of factory farming and meat consumption and the serious pull on grain demand this entailed, the post-war productivity boom in industrial farming was of such a magnitude that grain surpluses continued to far exceed domestic demand. And in the conflicting balance between supply restraint and conservation measures or export promotion as strategies to bring some stability to domestic prices, the policies of the New Deal and increasingly those which followed were more focused on export promotion in tune with agribusiness interests. Some degree of surplus US grain production had previously been exported to Europe, especially after 1870, but it was not until after the Second World War that the US agricultural sector took on its deep, enduring export imperative – becoming 'addicted to foreign market expansion' as 'a vital necessity to manage the tendency of capital accumulation to break down under the stresses of overproduction' (Berlan 1991: 119, 128).

After the war, outlets for increasing US surpluses were initially met by demand from a devastated Europe and paid for in part through the US-funded Marshall Plan. European agricultural sectors rebounded quickly from wartime devastation, however, and governments there soon came to view cheap food grain imports as a barrier to food security, economic growth and independence, establishing support programmes and very protectionist policies with the notable exclusion of feed crops, which were integrated into its rapidly industrializing grain-livestock complex. This forced the USA to look elsewhere for surplus outlets, and elsewhere quickly became the decolonizing South. In 1954, the US government established Public Law 480 (PL 480), the Agricultural Trade Development and Assistance Act (also known as 'Food for Peace'), which was subsequently moulded by the strong lobby of grain-trading and processing companies (Kneen 1995; Morgan 1980).

With PL 480, the USA institutionalized food aid and concessional sales to the developing world as the primary means for dumping its grain surpluses (with the latter, the recipient government typically received commodities below market cost and would in turn resell them for local currency, with the USA often making provisions as to how the funds could be employed). Food aid and concessional sales were variously an extension of domestic agricultural policy, a means to establish new markets for future commercial sales, as would soon

65

become more evident, and a foreign policy tool during the cold war with geostrategic objectives (Kodras 1993; Garst and Barry 1990). The latter motive is encapsulated in an oft-cited statement by US Senator Hubert H. Humphrey to the Agriculture and Forest Committee of the US Senate in 1957 that 'if you are looking for a way to get people to lean on you and to be dependent on you, in terms of their cooperation with you, it seems to me that food dependence would be terrific'. For the developing-country recipients of this cheap food, temperate grains came to displace or rival traditional dietary staples, discussed in Chapter 3, and this dependence forged by aid turned into markets for commercial trade over time. While food grains were the foundation of rising global agricultural trade, other elements of the grain-livestock complex were also increasingly exported over time, namely feed crops, dairy products and meat. The shift of US food aid into trade in the early 1970s was marked by a number of events. In 1971, the US trade balance slipped into deficit, where it has remained ever since, and the flagging industrial competitiveness coupled with the oil price shock of 1972–73 and the weight of its spiralling military expenditure on Vietnam and the cold war together gave an added impetus for increasing agricultural exports. The USA began selling grain to the Soviet Union in the early 1970s, causing temporary shortages and price spikes on world grain markets, and in 1973 the US Farm Bill explicitly made commercial export growth an objective of agricultural policy rather than being couched as a necessary by-product of domestic price stabilization. With this, the modest supply control policies that had existed were eliminated, a change welcomed by agro-food TNCs (McMichael 2004b; Friedmann 1993).

European grain-livestock exports had also begun to grow significantly around this time. Food self-sufficiency had been the major priority when the European Economic Community (EEC) established the Common Market in 1957, a goal that continued into the 1960s as the Common Agricultural Policy (CAP) put guaranteed prices and protectionism against imports in place for EEC-produced commodities. But the EEC's price supports together with the technological advances in farming did not only encourage aggregate self-sufficiency; as in the USA, large grain-livestock surpluses emerged, seen most vividly in the CAP's famous warehoused mountains of butter. As surplus production grew, led by France, the EEC also devised a variety of export support programmes to facilitate their dumping of grains, meat and dairy products.

With the declining priority of production controls and conservation in farm support programmes and expanding exports increasingly treated as the way to manage temperate grain and livestock surpluses, the competition for export turf in growing world markets intensified, particularly between the USA and the EEC (and its later expanded forms, the European Community/Union, EC/EU). With price supports in the EU and production payments in the USA giving incentives for expansion that overrode market logic (witness the large and protracted declines in world market prices), at the same time as technology and rising inputs were enhancing yields, a spiral was created whereby chronic grain surpluses brought pressures for more intervention and export growth, the nature of the subsidy regimes ensured that large surpluses would be exported and levels of state support for agriculture continued growing. In this export competition, state supports took a variety of different forms (e.g. export subsidies, export credits and state marketing programmes) while the strategic importance of agriculture shielded it from any multilateral policy commitments as agricultural funding continued to expand into the 1980s. For instance, greater direct export subsidies in Europe produced counter-initiatives like the US Export Enhancement Program, whereby the government depressed the export price to assist exporters moving into new markets.

Amid these escalating subsidies there is no indication that any thought was given to the dislocating impact that the externally dispersed surpluses might have on farmers in the recipient countries. Nor were the subsidy regimes designed with concern for their uneven internal distributional impacts, which have tended to mirror and exacerbate the polarizing dynamics associated with the industrialization and corporatization of agriculture. An OECD paper in the late 1980s concluded that it was 'difficult to detect major benefits for small farmers from most existing policy packages' in most of the world's industrialized countries (Winters 1988: 2) (Japan, South Korea and Taiwan are notable exceptions with much smaller average farm sizes, discussed in the following chapter, with Japan both a heavy subsidizer and a large net food importer). This pattern has only deepened since, most glaringly with US and EU subsidy regimes that have consistently rewarded scale, with the largest and most productive farms receiving the lion's share of the payments and very little reaching the bottom end of the farming spectrum.

For agro-food TNCs, the ideal model for subsidy regimes is to have minimal production controls, limited or no state intervention in prices (i.e. decoupling supports from price floors at the farm gate) and state subsidies concentrated on direct payments linked to the volume of production as compensation for depressed prices. This lets markets in chronic surplus set the prices, ensuring a cheap supply, while the volume-based compensation payments ensure that very large producers remain profitable and able to grow even if market prices dip close to or below the actual costs of production. The subsidy regimes in the USA and the EU have steadily approached this model with the USA in the lead and the EU moving sharply in this direction with its major CAP reform in 1993, which cut guaranteed prices and shifted subsidies to direct payments, in effect synchronizing 'EU policy with that of the US in favouring traders over producers' (McMichael 2004b: 6). Soon after, the USA extended this model again with the speciously named Freedom to Farm Act (1996) which further undermined supply management efforts (e.g. conservation set-asides) and did away with price floors for farmers, intensifying the spiral of falling prices, reflexive overproduction and rising, exceptionally uneven subsidy payments (Rosset 2006).

For the OECD as a whole, roughly 80 per cent of total farm subsidies are paid to the top 20 per cent of farmers as measured by farm income, a ratio that holds for both the USA and the EU. Within this top 20 per cent the unevenness grows further as the largest subsidy recipients tend to be the biggest landholdings, at the pinnacle of which have been the payments made to some extraordinarily wealthy individuals and massive corporate-owned farms in recent years (Economist 2005). The Institute for Agriculture and Trade Policy (IATP) and the Environmental Working Group (EWG) have both provided valuable detailed information about the imbalances of the US subsidy regime,[4] and even mainstream corporate media such as the *New York Times* (NYT 2003; Becker 2002, 2001; Weiner 1999) and *The Economist* (2005) have regularly criticized the skewed distribution of agro-subsidies. Because these subsidies constitute such a large share of the total net income in agriculture in the USA and the EU they have become a major point of contention in the struggle over the uneven multilateral regulation of agriculture, discussed in Chapter 4, with the perversions of existing subsidy regimes threatening to obscure the importance of state support for agriculture in a general sense. Meanwhile, the pressures associated with the chronic state of surplus have continued to be increasingly projected outwards on to foreign markets, as the waistlines of the US population can only expand so far. In the late 1990s the US Secretary of Agriculture described 'the expansion of global market opportunities for US agricultural producers' as being 'one of the US Department of Agriculture's [USDA] primary objectives', pursued through a series of 'aggressive efforts ... to bolster export competitiveness, open new markets, and expand exports' (Glickman 1998a).[5] By this time, exports had risen to 30 per cent of total agricultural receipts, up from 20 per cent in the mid-1980s, making agriculture twice as dependent on foreign trade as the US economy as a whole. As a US trade negotiator explained it, 'there is no other sector of the [US] economy where the link between trade and today's prosperity is clearer than in agriculture ... Exports are critical to nearly every sector of US agriculture' (Scher 1998). The significance of the agricultural trade surplus for the USA is magnified by its large overall trade deficits; a US senator insisted that the USA 'depend[s] on farm exports to keep our balance of trade sort of halfway in balance' (Bumpers 1998). While this last point is exaggerated, as agriculture barely dents US manufacturing deficits, such comments make plain the fact that domestic supply management is entirely off the radar of agricultural policy.

The most obvious trend in the structure of US agro-exports is the stagnation by volume and decreasing relative importance of grain and feed exports, decreasing from 46 to 27 per cent of total exports over the last two decades of the twentieth century, along with the rising volume and relative importance of meat and livestock products in the overall agro-export structure, increasing from 9 to 20 per cent of total exports during this time. US meat production increased by roughly half in the last two decades of the twentieth century, and the eightfold increase in meat exports accounted for one-third of this growth. Fruit and vegetables have also become an increasingly significant part of the US agro-export structure, growing from 6 to 14 per cent of total exports over this same period (USCB 1999). As the then Secretary of Agriculture explained, since the 1980s 'the big increases in [US] exports have been in value-added, which is livestock, meat, poultry, and fresh fruits and vegetables, and those have been dramatic, profound increases', and these are where the USDA has identified 'great positive potential' for further export growth (Glickman 1998a). Related to this potential, as officials have also noted, is the need to shift dietary patterns in line with US exports, with fast-food restaurants recognized as having important roles (Scher 1998).

The desire for further export growth was a major reason why agriculture was included in both the WTO and the North American Free Trade Agreement (NAFTA) negotiations (the latter between the USA, Canada and Mexico), and why the USDA described both as 'landmark' victories (Glickman 1998b; Schumacher 1998; Scher 1998). While US agro-exports subsequently expanded, however, so too have its agro-imports, to such an extent that the US agricultural trade surplus has actually shrunk considerably in recent years, an outcome that helps to bring the real victors of institutionalized trade liberalization into plain view: it is not US agriculture per se but rather the TNCs which have been dominant, behind-the-scenes actors directing the course of multilateral trade regulation. To understand further how these corporations are the dominant interests giving shape to and benefiting from multilateral trade regulation – explored further in Chapter 4 – their tremendous power within the agro-industrial complexes of the temperate world must first be examined.

Agro-giants: the ascendancy of transnational corporations

The growth and concentration of corporate control have steadily increased in the global economy to the point where the top two hundred TNCs control the equivalent of 29 per cent of global economic activity, the total sales of the ten largest TNCs exceed the total GNP of the world's hundred smallest national economies and, when measured together with nation-states, individual TNCs account for half of the world's hundred largest economies. The extensive spatial and biological character of farming was once seen to have partially inhibited the monopolizing tendencies of capital over agriculture relative to other economic sectors, but agro-TNCs have steadily overcome these barriers to gain control over both the input and output sides of farming, or upstream and downstream of the land, as framed by Bernstein at the outset of this chapter. While government programmes and market intervention have continued to play an important role in key realms of the temperate grain-livestock complex it is the large, growing and consolidating corporations which have clearly become the dominant actors integrating the agricultural system from farm to market, especially from the 1970s onwards, with the USA again at the forefront of these changes.

Upstream: sowing input dependence The ability of capital to appropriate surplus from agriculture was impeded by the fact that production is based on large numbers of spread-out, independent farmers whose ability to produce in closed-loop agro-ecosystems was – aside from access to enclosed land itself – until recently mostly beyond the control of the market (Lewontin 2000). As noted, the transformation of farms into through-flow systems dependent upon external inputs created opportunities for capital to control and profit from the production side of agriculture. The commodification of the seed was simultaneously a direct source of profits (a form of technological rent that was contingent on intellectual property protections), a means of securing profits with other inputs and a means of protecting the system against reversal, because if the seed could be transformed into a commodity its owners would be put in a commanding position to technologically weave together dependence upon seasonally purchased seeds with dependence upon other inputs, which would prevent farmers reclosing agro-ecological loops and escaping this off-farm purchase and control. Kloppenburg (2004: 201) skilfully summarizes this power of the seed:

> ... a seed is, in essence, a packet of genetic information, an envelope containing a DNA message. In that message are encoded the templates for the subsequent development of the mature plant. The content of the code crucially shapes the manner in which the growing plant responds to its environment. Insofar as biotechnology permits scientific and detailed 'reprogramming' of the genetic code, *the seed, as embodied information, becomes the nexus of control over the determination and shape of the entire crop production process* [emphasis in original].

Thus, agro-input TNCs have pursued the commodification of seeds with great vigour. But while the prize is great, it is a formidable challenge.

As both biologically regenerative entities and the outcomes of diffuse, long-term selective breeding efforts, held as a collective inheritance by many people and cultures, seeds are a particularly resilient sort of common property (ibid.; Shiva 1993). Through the ages farmers have saved, selected and shared seeds, and even when they might purchase seeds in the market this need not have entailed a continual dependence on such purchases, nor would the seller have any legal right over the nature of the seed itself. In other words, if a farmer purchased seeds in a market one year, they could save the seeds from the crop and sell them in the same market the next. In order to fully commodify the seed a life form itself must be considered to be private property and hence patentable – which until quite recently was a legal impossibility and which remains offensive and immoral to many people and cultures. It also demands that there can be private or intellectual property rights over scientific knowledge and innovations which are enforceable through patents and associated legal protections, in ways that reify modern manipulations and their agents and discount the history of seed selection and its multitudinous inheritors (Shiva 1993). Conversely, the FAO (1997: 19) describes how 'farmers' varieties, otherwise known as landraces or traditional varieties' should be understood as 'the product of breeding or selection carried out by farmers, either deliberately or not, continuously over many generations', while noting how traditional seed diversity is not easily patented since farmers' 'varieties tend not to be genetically uniform and contain high levels of genetic diversity' and therefore tend to be 'difficult to define or distinguish unequivocally as a particular variety'.

While much of the initial research into enhanced seeds was conducted by public agencies or quasi-public international institutions, which still have an ongoing role in this process, agro-chemical corporations began progressively seizing control of the direction of seed enhancements and the ownership of new varieties with a wave of buying up smaller, localized seed companies around the world, and extensively integrating the agro-input sector vertically (chemical–seed–fertilizer–pharmaceutical) and horizontally. At first glance the integration with pharmaceuticals might seem to be more indistinct, but it comes into focus in view of the scale of antibiotic and hormone usage in industrial livestock rearing. In the USA, livestock consume eight times more antibiotics by volume than the human population, which, as noted, permits the scale of confinement and abets rapid weight gain. Between 1985 and 2000 the *average* sub-therapeutic volume of antibiotics used in farm animals in the USA rose by 50 per cent and tripled per poultry bird (Nierenberg 2005; Mellon et al. 2001). In the quest to commodify the seed, Cummings (2005) describes maize as the 'Holy Grail' because of its centrality to industrial agriculture, with innovations in hybrid maize in the 1950s a major early step binding farmers into the annual purchase of seeds. Soy soon became another key target in the corporate control of seeds, and today the seed market in the USA for these two core feed crops alone is valued at US$5 billion, with two companies (Monsanto and DuPont Pioneer) controlling roughly 60 per cent of the market (Hendrickson and Heffernan 2005).

The agro-input industry is today a complex oligopolistic web characterized by growth, mergers, takeovers, joint ventures, interwoven licensing agreements and the continual striving for market expansion. In 2004, the top ten TNCs controlled 84 per cent of the US$35 billion global agro-chemical market (led by Bayer, Syngenta, BASF, Dow, Monsanto and DuPont); the top ten TNCs controlled roughly half of the global US$21 global commercial seed market (led by Monsanto, DuPont/Pioneer Hi-Bred and Syngenta); and the top ten TNCs controlled 55 per cent of the US$20 billion global animal pharmaceutical market, the greater part of which is used in animal agriculture (ETC Group 2005a, b). This consolidation is widely expected to continue deepening in the coming years, in turn magnifying the costs of input dependency for industrialized farmers. This input dependency is such that farmers are ensnared not only by interlocking inputs at the point of purchase but by the hierarchical system of agricultural research wherein they are treated as recipients of research rather than agents or co-participants. And as companies vertically integrate and weave together inputs in a through-flow industrial system, and farmers' understanding of closed-loop agro-ecosystem functioning is correspondingly eroded, one crucial outcome is that where farmers used to go to other farmers for advice they are increasingly forced to turn to input suppliers and chemical companies, with the shifting of this knowledge base a key mechanism for surplus extraction from farmers over the long term. Further, the consolidating control on the output side of farming also interacts with the pressures for input standardization as the narrowing control over farm output has also led to strict demands for the size, appearance and timing of farm production, factors that contribute to the systemic pressures to use agro-input packages (Lyson and Lewis Raymer 2000; Heffernan 2000; Heffernan et al. 1999).

Since the 1980s, a major priority for agro-input TNCs in extending the technological dependence of farmers has been the development and commercialization of GM crop varieties, something which is 'widely recognized as one of the most important chapters in the recent development of global agriculture' (Buttel and Hirata 2003: 1). In the 1970s some large agro-chemical corporations in the USA began experimenting with research that involved snipping and embedding genes across species barriers using recombinant DNA technology, which ultimately succeeded in producing commercially viable GMOs (Cummings 2005). To critics, GM crops raise weighty ethical questions about the ability of humans to shape nature on a fundamentally new level and magnify the ecological risks associated with the simplification and toxicity of industrial agro-ecosystems. Some of the major ecological risks include unpredictable outcomes associated with genetic transfers from GMOs to crop relatives and wild plants (sometimes called 'genetic pollution'), the possibility that new viruses will emerge and the likelihood that new pest and weed resistances will develop over time, with the hazard of 'super-weeds' often stressed, since so much GMO research has focused on engineering herbicide resistance – all of which inescapably punishes any nearby organic farmers by precluding their ability to certify (ibid.; Kloppenburg 2004; Middendorf et al. 2000; Altieri 1998).

To counter such criticism, advocates of GMOs typically hold up the scale of global hunger and undernourishment, the fact that the human population will grow by roughly 3 billion people in the next two generations and projections of continued increases in total and per capita global demand for meat, along with the topping out of the productivity gains of high-input monocultures, and against this backdrop then frame GMOs as the 'next Green Revolution' of higher-yielding crops which is urgently needed to keep pace with rising demand and ultimately avert global food shortages. There is also the claim that GMOs could provide a cure for widespread health problems such as micro-nutrient deficiencies in poor countries (e.g. 'Golden Rice'). At the extreme end of the celebratory and humanitarian claims for GMOs is the accusation that opposition is actually perpetrating an injustice upon the world's poor by depriving them of these potential breakthroughs, which, much as occurred with the Green Revolution, intentionally obscures the distributional problems with both production and consumption (McMichael 2004a; Friedmann 2000).

Chrispeels (2000: 3), a scientist in the pro-GMO camp, calls the argument that distributional issues are at the heart of global hunger issues 'as facile as it is incorrect', without any explanation other than a conversely simplified assertion that 'food production and purchasing power both need to increase in developing countries' and 'food production needs to increase in developed countries as well so that grain can be exported at a price the poor can afford'. Further, in a fabulous twist of the economic logic of seed commodification, he argues that 'since the cost of these inventions is being charged to the consumers in the developed countries, such approaches amount to a transfer of wealth by large corporations from the developed world to the developing world', and claims that the 'ethical considerations of genetic engineering of crops pale in comparisons to the ethical considerations of not improving the lives of the poor' (ibid.: 5–6). Other GMO advocates attempt to frame the ethical debate in similar ways, as though it is between waiting for 'conclusive evidence' versus 'stalling the potential benefits for so many in need', and go so far as to project the case for GMOs into the voice and moral appeals of the poor, insisting that 'there is a strong argument, *from impoverished communities especially*, to facilitate and bring about advancements in GM crops to not only increase trade but to improve health and nutrition' (Von Braun and Brown 2003: 1044; emphasis added). When the case for GMOs is framed in such ways, reasoned appeals for the precautionary principle, transparency, further research into environmental and health implications and public debate on alternatives and ethical issues often end up being dismissed as 'emotional' and 'unscientific' barriers to progress.

But the humanitarian claims for GMOs grow dim when two of the major foci of GM research are examined and we see how the so-called seed 'enhancement' has nothing to do with improving the human condition and everything to do with enhancing the scope of surplus extraction. First, the predominant 'enhanced' trait present on GMO cropland today is a biologically encoded compatibility with a specific agro-chemical, mostly herbicides, thus locking farmers into the purchase of seed and chemical packages (Buttel and Hirata 2003; Shiva 1999). The most famous case is the seeds engineered by Monsanto to be compatible with its broad-spectrum herbicide Roundup, with Roundup Ready seeds available in such crops as maize, soy, canola/rapeseed and cotton. The second case is the research effort devoted to the design of 'terminator seeds', so named because a toxin is inserted into a gene in order to sterilize the plant. Though its commercial application has so far been blocked, terminator research sheds remarkable clarity on the central motivation behind the GMO research agenda. Even the most acrobatic rhetorical gymnastics cannot connect terminator seeds to any moral justification, though an attempt has been made through a contradictory line of reasoning that invokes the otherwise denied or downplayed danger of genetic pollution from GMOs and then argues that the terminator would eliminate such concerns (Dundon 2003). But the real motive behind the terminator seed for corporations is unmistakable: it would shift the *seed as commodity* from a more tenuous *scientific-legal conception*, where it can be contested in various ways (e.g. saving seeds, challenging patents), to a *biophysical attribute* whereby their annual purchase is simply irresistible. This motive is hardly concealed in the industry's own euphemistic labelling, calling this a 'Technology Protection System'. Monsanto has been the leading actor here, and the USDA has also advanced terminator research and framed it as a tool for fostering market growth abroad (Lewontin 2000; Middendorf et al. 2000). Shiva (1999: 36) summarizes the essence of this research agenda: 'Termination of germination is a means for capital accumulation and market expansion ... When we sow seed, we pray, "May this seed be exhaustless". Monsanto and the USDA, on the other hand, are stating, "Let this seed be terminated so that our profits and monopoly is [*sic*] exhaustless."'

Yet however they attempt to justify GMOs, the agro-input TNCs driving this research and its commercialization have not been interested in public discussion on the matter. Rather, they made a clear strategic effort to get 'to the lawmakers before the public did', given their acute awareness of the increased regulations that had followed the research of Rachel Carson and other environmental scientists and the associated public outcry (Cummings 2005: 27). And they were especially successful in this effort in the USA, coming to dominate the government institutions responsible for agricultural regulation and research: the Secretary of Agriculture and the USDA, the US Food and Drug Administration (FDA) and the extensive system of land grant universities (ibid.; Mattera 2004; Buttel 2003; Dundon 2003). Mattera (2004) gives a detailed account of how the USDA, dubbed 'the people's department' at its inception by President Lincoln in 1862, given the large population of family farmers it was created to serve, was transformed into 'Agribusiness Industry's Department' for its role as a permissive regulator and in subsidizing corporate research and extension agendas. A major aspect of this is the revolving door between senior USDA appointees and executives, lobbyists and lawyers who once worked or who continue to have an association with agro-input TNCs, which routinely opens into policy-making positions in government regulatory bodies and back out into lucrative corporate jobs, while being almost entirely closed to the interests of family farmers, agricultural labourers, consumer rights and public health advocates and environmental organizations. This reached its apogee under the George W. Bush administrations, the recipient of soaring contributions by agro-TNCs, with the most obvious cases being former Secretary of Agriculture Anne Veneman, previously an executive of the chemical and biotech company Calgene (later taken over as a Monsanto subsidiary), and Secretary of Defence and former über-insider Donald Rumsfeld (who previously headed Monsanto pharmaceutical subsidiary Searle), but which also included the stacking of other key regulatory agencies and senior aides with former Monsanto employees and people who have long worked for large agro-TNCs and their lobby organizations (ibid.). The US FDA has also been greatly influenced by pro-biotech planners and administrators who have come from agro-input TNCs, capped by a former Monsanto lawyer taking charge of FDA policy development. Commercial agriculture divisions of the US land grant universities extend further this takeover of ostensibly public institutions, led by the University of California in the research on GMOs (Cummings 2005). Another very notable aspect of the corporate–state nexus is in the realm of trade negotiations, as will be explored in Chapter 4.

In short, agro-input TNCs have been able to craft an exceptionally supportive regulatory environment that has allowed them to unleash GMOs upon the US landscape in almost complete silence. This supportive regulatory environment includes muscular legal protections for intellectual property rights, which allow GMO patents to straddle a tricky legal and conceptual dichotomy: on one hand patent claims rest on demonstrating that an innovation is different enough to represent intellectual property but on the other, to allay environmental and food safety concerns, there is a simultaneous assertion of substantive similarity or equivalence to traditional crops. It also includes the sanctioning of commercial usage far ahead of the understanding of long-term environmental or health impacts with no rigorous safety studies, post-market monitoring or independent reviews even required and an absence of laws that demand consumer labels for GM foods (ibid.). Beyond the USA, agro-input TNCs have also been successful in largely suppressing public discussion about GMOs in a few other of the world's major industrial grain- and livestock-producing countries, principally Argentina, Canada and China, where GMOs have also become a significant part of agricultural landscapes (Buttel and Hirata 2003).

Once established in patent rights and on fields GMOs can also entrap disinclined farmers in nefarious ways, through the uncontrollable drift of seeds and inadvertent genetic pollution of non-GM fields coupled with the sampling and legal vigilance of proprietary agro-input TNCs. The case of Canadian canola/rapeseed farmer Percy Schmeiser versus Monsanto highlights this dramatically. In 1998, Monsanto filed a patent infringement suit against Schmeiser, alleging that he stole Monsanto seeds because the company's investigative scientists found their genetically encoded crops on his fields when he had not purchased a licence. Schmeiser countered by asserting that he had not planted Monsanto seed varieties and that Monsanto was the guilty party for genetically polluting his land, but the Supreme Court of Canada ruled that irrespective of how GMO seeds get into a farmer's field they are the company's property. For Monsanto this was an auspicious precedent, having so much invested in GMOs both in research costs and in how perceptions of market growth potential play out for investors and hence stock prices. In aggressive legal manoeuvring to enforce patents (Goldsmith 2004; Schmeiser 2004)[6] we see specialized patent lawyers emerging as a new and pungent face of industrialized agrarian change today.

While GM seeds have already been deeply established in a few crops (maize and soy) and a few countries, however, it is important to emphasize that in the rest of the world – and very notably Europe, which along with the USA is also home to many of the world's biggest agro-input and agro-food TNCs – vigorous public debate and opposition have resisted this new technology. This has made GMOs one of the great faultlines in multilateral efforts to institutionalize the trajectory of the global food economy, discussed in Chapter 4.

Downstream: diminishing farm-gate earnings Agro-food TNCs have also squeezed the surplus in agriculture from farmers on the output end, reflected in a rising share of the average food dollar being contained in processing, distribution and retail. Apart from the huge grain traders that had older global ambitions (Morgan 1980) and a few trademark processed goods (e.g. Coca-Cola), most of the large US food and beverage companies had been primarily focused on regional and national markets into the 1970s. The technological compression of time and space and accelerating vertical (processing–distribution–retail) and horizontal amalgamations, however, enabled large national companies to become increasingly international in orientation, with their subsequent growth in global markets 'nothing short of phenomenal' (Lyson and Lewis Raymer 2000: 202; Heffernan 2000; Lehman and Krebs 1996; Sexton 1996; Heffernan and Constance 1994; Friedmann 1993; Krebs 1992). Agro-food TNCs have been able to cross-subsidize their operations by operating in multiple sub-sectors, using profits on one product to temporarily absorb losses on another, which has enhanced their ability to penetrate new markets by lowering prices and eliminating local competition (Heffernan 2000). They have also proved remarkably adept at changing dietary aspirations and habits, building strong brand loyalty into food choices through advertising efforts while simultaneously making the food economy less visible, less intelligible and less concern-worthy to consumers, as well as ceaselessly finding opportunities for value-added (but nutrition-deprived) products in the time pressures of hectic modern lifestyles, from fast food to ready-made meals to junk-food snacks.

Perhaps the ultimate reflection of 'branded' loyalties and despatialized conceptions of food at the heart of the global food economy is the proliferation of fast-food chains – the annual fast-food expenditure in the USA rose from US$6 billion in 1970 to US$110 billion in 2001 (Schlosser 2002). The transformation in how food is understood is also plainly illustrated in the fact that by the late 1980s only about 8 per cent of the products found in US supermarkets were not processed (George 1990) and that Wal-Mart, the world's largest general retailer, become the largest grocery retailer in the USA as well as on a global level in the 1990s. Though it had barely begun to compete in the agro-food market less than a decade earlier, by 2003 Wal-Mart's grocery sales had grown to US$66 billion in the USA and US$245 billion worldwide (Hendrickson and Heffernan 2005), a magnitude such that it is now understood to be 'forcing many changes in retailing at the global level' (Hendrickson et al. 2001: 3).

The scale and pace of consolidation in agro-food TNCs is illuminated in research by rural sociologists at the University of Missouri for the National Farmers' Union and by that of the Action Group on Erosion, Technology, and Control (Hendrickson and Heffernan 2005, 2002; Hendrickson et al. 2001; Heffernan et al. 1999; ETC Group 2005b). In the USA, the top four corporations in grain milling – Cargill, Archer Daniels Midland (ADM), ConAgra and Cereal Food Producers – account for 63 per cent of the national total, up from 40 per cent in 1982; the top four corporations in soybean crushing (ADM, Bunge, Cargill, Ag Processing Inc.) account for 71 per cent of the national total, up from 54 per cent in 1977; and the top three corporations control 82 per cent of all maize exports (Hendrickson and Heffernan 2005). In short, there are a small number of firms controlling the US cereal and feed crop industries, and these have been the key beneficiaries of the post-war export promotion strategies that carved new import dependencies, as aid and concessional sales turned into commercial trade from the 1970s onwards and as the subsidy regimes, surplus production and long-term farm-price declines have continued fortifying their international competitiveness.

The beef and dairy, pig meat, poultry meat and egg industries have also experienced rapid growth and consolidation with a partially overlapping group of dominant corporations in livestock slaughter and processing. In the contract growing arrangements pioneered by the broiler industry, Tyson Foods and other large meat producers control and integrate the feed mills, chick hatcheries, slaughter, processing and retail distribution, but prefer to outsource the warehousing and growing of animals to farmers, who can more aptly be called contract growers. In this system, the slaughter-packer-distributor provides the chicks and technically owns the animals while transferring the cost of the factory farm unit itself, its depreciation and upkeep (including company-mandated building upgrades), the risk of growing, the risk and cost of waste management (not only of the immense amount of manure produced, but of the significant number of animal carcasses which do not make it to slaughter) and even the transport costs to the grower, while continuing to stipulate and sell the feed and pharmaceuticals. Broiler flocks are generally turned over in six to eight weeks, and with a few large, integrated companies controlling access to the market (with the oligopolistic character of the industry even more acute for farmers at a local scale) growers are trapped in what has been called a 'captive supply' with very little leverage to affect the nature of the contract and very small margins for each animal. Contract arrangements are also fast growing with factory-farm pig production, and the increasing integration between beef feedlots and slaughter and packing plants has led to different forms of contracts with farmers (Nierenberg 2005; Midkiff 2004).

In the 'big three' of animal production in the USA, the top four corporations alone control a large and fast-rising market share: in broiler chicken slaughter and packing, the top four corporations (Tyson Foods, Pilgrim's Pride, Gold Kist and Perdue) control 56 per cent of the industry, up from 35 per cent in 1986; in pork slaughter and packing, the top four corporations (Smithfield Foods, Tyson Foods, Swift and Co. and Hormel Foods) control 64 per cent of the US total, up from 37 per cent in 1987; and in beef slaughter and packing, the top four corporations (Tyson Foods, Cargill, Swift and Co. and National Beef Packing) control 84 per cent of the US total, up from 72 per cent in 1990 (Hendrickson and Heffernan 2005). Given the rapid rate of growth in the poultry industry it is not surprising that the biggest poultry producer, Tyson Foods, recently took over Iowa Beef Producers, which had the biggest beef packer and second-biggest pork packer in the USA. Tyson is now the world's biggest meat producer with sales and operations in Argentina, Brazil, China, India, Indonesia, Japan, Mexico, the Netherlands, the Philippines, Russia, Spain, the UK and Venezuela. In 2000, Tyson alone slaughtered 2.2 billion chickens (Nierenberg 2005).

Through horizontal integration, the largest agro-food TNCs now have exceptionally diverse tentacles, although this is partly clouded by the profusion of different brands, which sometimes mark the parent but often do not. The top ten agro-food TNCs in the USA had 2003 sales that topped US$142 billion: Altria (Philip Morris/Kraft Foods) ($22 billion), Tyson Foods ($22 billion), Pepsico ($18 billion), Con-Agra ($17 billion), Nestlé (US division) ($14 billion), Anheuser-Busch ($11 billion), Mars ($10 billion), Sara Lee ($10 billion), General Mills ($10 billion) and Dean Foods ($9 billion) (Hendrickson and Heffernan 2005). Many of these same corporations appear on the list of the world's top ten agro-food TNCs in terms of 2004 sales: Nestlé ($64 billion), ADM ($36 billion), Altria ($32 billion), PepsiCo ($29 billion), Unilever ($29 billion), Tyson ($26 billion), Cargill ($24 billion), Coca-Cola ($22 billion), Mars ($15 billion) and Groupe Danone ($17 billion) (ETC Group 2005b). As discussed in Chapter 1, the opposite poles of hunger and undernourishment and rampant obesity and diet-related diseases are an obvious indication of the contradictions of the global food economy, and this group of giant agro-food TNCs purveys a wide array of unhealthy but highly profitable junk foods. Thus, it is only logical that in 2006 Nestlé, the world's biggest agro-food TNC and also a leader in junk food, purchased the Jenny Craig franchise, a major brand in a weight-loss industry whose total value is estimated at US$50 billion (NYT 2006).

It is difficult to assess the relative balance of power between food processors and food retailers with such large and multi-faceted agro-food TNCs interacting with large and sometimes massive (e.g. Wal-Mart, Carrefour, Tesco, Metro and Kroger) retail chains. Most people in the industrialized world and increasing numbers of the middle and upper classes in the developing world purchase large amounts of their food at supermarkets, putting large retailers in an influential position to set the terms with food manufacturers. Conversely, however, 'retailers are at the mercy of those manufacturers who have successful brands because branding is one way to create leverage with retailers', though whoever holds the upper hand, 'the basic point is that smaller entities in any part of the chain are being left out, and that processors are much more focused on serving the interests of food retailers rather than the interests of farmers' (Hendrickson et al. 2001: 12–13).

In sum, the global food economy is becoming progressively less democratic as agro-input and agro-food TNCs appropriate a growing share of both control and surplus in agriculture. In the case of the US food economy, which has global implications, the trajectory is framed well by Heffernan et al. (1999: 13–15):

> … the major decisions in the food system are being made by an ever-declining number of firms, a growing number of which are involved in the food system clusters … [this] centralized food system that continues to emerge was never voted on by the people of [the USA], or for that matter, the people of the world. It is the product of deliberate decisions made by a very few powerful human actors. This is not the only system that could emerge. Is it not time to ask some critical questions about our food system and about what is in the best interest of this and future generations?

The polarity of production in industrial agriculture

Farm operators Since 1985, a number of famous rock and country music stars have gathered for an annual fund-raiser in the US Midwest to support ailing American family farmers, an event known as Farm Aid. For small farmers in the developing world who are being squeezed by industrialized grain-livestock surplus exports, this must seem to be the most peculiar of events. Farm Aid is, however, but one sign of the fact that even in the world's most productive agricultural system, and the birthplace and headquarters of so many giant agro-TNCs giving shape to the global food economy, few farmers are actually benefiting from the current system.

In 1930, there were 7 million farmers in the USA, more than one-fifth of the nation's workforce, and the farm population including families was over 30 million, representing nearly a quarter of the nation's total. By 1970, little more than a generation later, the US farm population was only 4 per cent of the nation's total. Goldschmidt's (1947) oft-cited study in rural California comparing the structure of the agricultural system and the social health of rural communities provided both an empirical observation and a warning amid this transformation. Goldschmidt concluded that communities where small farms and small businesses still prevailed were much healthier than the communities where large farms were dominant and corporations were increasingly controlling farm inputs and outputs, as the surplus generated by agriculture circulated within the community in the former was leaking out in the latter. These patterns have played out time and again. Between 1910 and 1990, the share of the total value within the US agricultural system going to farmers fell from 41 to 9 per cent, while the share contained in inputs rose from 15 to 24 per cent and that in processing, distribution and retail increased from 44 to 67 per cent (Lehman and Krebs 1996). As Magdoff et al. (2000: 12) note, 'farming is one of the few businesses that pays retail prices for inputs and sells its products at wholesale prices', as farmers buy from commercial suppliers but rarely sell directly to the market. The 'double squeeze' of rising input costs and low and falling farm-gate prices reduced profits per unit area, with the eroding margins of this cost–price squeeze producing serious scale and mechanization imperatives and making smaller holdings less viable – the proverbial pressure to 'get big or get out'.

In order to access machinery, new inputs or expanded farm area in line with these pressures, farmers routinely financed operations out of equity or took on unsustainable debt loads. For those farmers who are on the cusp of the technological innovations or growing in land area, there can be a short-term economic gain from increased individual productivity at current price levels, but when innovations spread, output widely increases and prices fall, modernizing farmers are put back in the cost–price squeeze. Overextended farmers can be extremely vulnerable to rising interest rates, crop failures and falling or persistently low prices, which is why land sales, foreclosures and farmer suicides all became common features of the industrializing

US farm landscape. Such land, in turn, has tended to pass to big landowners or corporations that can better navigate the cost–price squeeze through scale, concessions (e.g. tax, inputs and credit) and the subsidy regime skewed heavily in their favour. This was especially dramatic in the immediate wake of the Freedom to Farm Act (1996), as average market prices plummeted and the number of farms of less than 2,000 acres (810 hectares) in size fell by more than 90,000 over the next five years, at the same time as the number of farms greater than 2,000 acres in size grew by more than 3,600 (Rosset 2006). Rosset (ibid.: 49) sums up the dynamics of the subsidy regime in this process: 'large growers are compensated with direct payments for producing at such low prices, while family farmers, the vast majority, get virtually nothing, and many are driven out of business'.

The net result has been a remarkable concentration of agricultural production and land and has made independent farming a vocation that is exceptionally forbidding to enter into and nearly impossible to re-enter once driven out. A few snapshots of the US farm economy at the end of the twentieth century indicate the magnitude of this transformation: the area of cultivated land per agricultural worker in the US (58 hectares) is roughly fifty times the world average (Gilland 2002); farms greater than 500 acres (203 hectares) control 79 per cent of all US farmland; a mere 8 per cent of farms (with sales of at least US$250,000) account for 72 per cent of all farm sales along with the majority of farm subsidies; and the population calling farming their principal occupation was under 1 million, with the bulk of production coming from only about one-third of these (USCB 1999). There are more prisoners in the USA than full-time farmers. The flipside of the massive industrial operations that provide the bulk of production are the much smaller-scale family farms barely hanging on. The National Family Farm Council notes that 73 per cent of US farms account for only 7 per cent of total farm sales. In 1990, more than one-fifth of all farm households were found to have had incomes below the official poverty line, twice the rate of all US households (Lehman and Krebs 1996), even before the precipitous decline in US farm income in the late 1990s. In Europe production has not become quite as asymmetrical as in the USA though similar trends have been at work in spite of greater controls on farm size. In France, Europe's leading agricultural nation and headquarters to some of its biggest agro-TNCs, today's farm population is only a quarter what it was in 1950 and polarization is picking up speed. Between 1993 and 2004 the number of farms in France fell by almost one-third and the average farm size grew by 40 per cent, though it is still less than a quarter that of the USA (Economist 2005). For Europe as a whole full-time agricultural employment fell by half between 1980 and 2001 and as many as 1 million farm jobs were lost in the second half of the 1990s alone (McMichael 2004b).

The move towards factory farming is another major part of the polarization of industrial agriculture, with the USA again at the vanguard of the process and hence both the most extreme and illustrative case as regards where the model is heading. In poultry and pig factory farms, the nature of contract farming has transformed the growers into essentially piecemeal workers for the vertically integrated companies. Because of the great start-up capital costs associated with a factory farm unit (roughly US$150,000 per broiler house and even more for layer houses) they are often established through debt – but the high risk, low margins and health, stench and aesthetic impacts of manure lagoons lead to steep depreciation, meaning that the grower is not building equity. In 1950 there were about 2 million pig farms in the USA, in the mid-1980s there were 600,000, and today the bulk of pig production comes from fewer than 74,000 operations. Huge volumes of the nation's beef cattle are fattened on a very small number of giant feedlots which are either owned or in contract with the firms responsible for slaughter and packing (Midkiff 2004).

Farm workers The unevenness of industrial agriculture becomes even greater when the lens shifts from the farm operators to farm workers. Majka and Majka (2000: 163) call farm workers 'a "super-exploited" segment of the US working class', with poverty levels nationally at a staggering 61 per cent in the mid-1990s and work contracts characterized by low wages, poor-quality and segregated living conditions and limited benefits, security and protection from toxic work environments. The US government estimates that at any given time there are 1 million immigrant farm workers at work in the farm economy, mainly from Latin America, of which roughly 40 per cent are undocumented. There is now open industry and government acknowledgement 'that, for better or worse, foreign-born workers are now one of the most vital elements in the American food and agriculture system' (Barboza 2001). The fact that there is a labour surplus among poor, recent immigrants undercuts the potential for organizing and collective bargaining, especially for those who are 'illegal' or who have family members with insecure status (Majka and Majka 2000).

Because extensive grain production and factory farms are so mechanized, relatively little labour is needed in these segments of the industrial grain-livestock complex; rather, the weight of human labour in the grain-livestock complex is centred on processing. Animal slaughter and processing is by far the most dangerous sub-sector of the US farm economy and, not surprisingly, it is also increasingly dependent upon an insecure and high turnover and heavily new immigrant labour force. Both compounding and partly because of this surplus cheap labour force, the strength of unions organizing in this sector has rapidly deteriorated. The US Department of Labour and the US Occupational Safety and Health Administration have both documented the extraordinarily hazardous conditions of slaughter and processing jobs with rates of serious accidents, maiming and repetitive stress and cumulative trauma injuries (e.g. carpal tunnel syndrome) far above national manufacturing averages, and in the USA these are compounded by extremely uneven access to healthcare coverage, especially for recent migrants (Human Rights Watch 2005; Midkiff 2004; Schlosser 2002; Barboza 2001; Stull et al. 1995).

A report on the poultry industry by the *Washington Post* described the slaughterhouse as 'a world defined almost exclusively by time and volume' with tremendous pressure to accelerate the pace of the kill floor, as 'the profit for chicken averages 2.5 to 3.5 cents per pound' and there is no regulation for the line speeds (Sun and Escobar 1999). An average US chicken slaughter and processing plant will kill and refine into packaged meat 150,000 chickens in a single shift – or 300,000 chickens every working day. The fast pace of the kill floor amid desperate animals and sharp blades brings great risk of injury for the workers, to say nothing of its implications for animal suffering, and it obviously produces equal pressures on the disassembling lines processing the bodies. The same *Washington Post* report noted how an average worker on a poultry breast processing line makes a cut through meat and bone almost every two seconds, or 1,600 times an hour (ibid.). A detailed survey of the working conditions in US slaughter and processing facilities by Human Rights Watch (2005: 1) summarized the system: 'the carcasses hurl along evisceration and disassembly lines as workers hurriedly saw and cut them at unprecedented volume and pace. What once were hundreds of head processed per day are now thousands; what were thousands are now tens of thousands per day.' It then went on to argue that the inordinate risk of injury and chronic ailment, insecurity of work, threats to organizing and lack of legal protection and recourse in slaughter and processing plants are not merely difficult working conditions but constitute 'systematic human rights violations' (ibid.: 2; see also Midkiff 2004; Schlosser 2002; Stull et al. 1995). Workers in factory farms face constant exposure to dust and the toxicity from the faecal matter, bacteria and pesticides, heightening risk of acute and chronic respiratory disease, but the number of people affected and the risk of sudden injury in factory farms are considerably less than in slaughter and processing plants. The one dangerous and labour-intensive job in poultry factory farms occurs when the birds reach slaughter weight and 'catchers' are sent in to round up the birds by hand, as efforts to mechanize this have so far failed, crushing or bruising too many birds (Sun and Escobar 1999).

Some accounts that focus on the terrible conditions of working people in slaughter and processing plants and factory farms stop here. Others, however, insist that this is only one aspect of the matter, emphasizing the great emotional and psychological burden of jobs whose repetitive focus on the death and dismemberment of sentient beings is marked by job titles such as knockers, stunners, hangers, liver pullers, navel boners, gizzard cutters, thigh-bone poppers and lung gunners (Silverstein 1999). Eight-hour shifts immersed in killing animals, hearing their cries, sensing their palpable terror and slicing their carcasses can 'desensitize and to an extent even dehumanize the workers', with the mental health and stress inevitably carrying 'over to other aspects of their lives' (Midkiff 2004: 36). At a poultry processing plant, a hanger might fasten as many as fifty birds a minute by their feet on to the swift-moving line where the bird's throats will be cut while they are writhing and fully conscious. And because of the speed of the lines and the movement of the birds the slaughter is sometimes only partial, so that wounded birds continue along the line in pain until they are killed by immersion in scalding-hot water. Similarly, 'humane' slaughter techniques for pigs and cattle also routinely fail, with improper pig knocking or bovine stunning sending many animals to their slaughter fully conscious (Mason and Singer 1990).

Clearly the industrial grain-livestock complex has created a fast-expanding realm of violence that exists outside of either widespread societal awareness or normative moral concerns. Further, the factory-farm methods that have been pioneered in temperate, industrialized countries are now fast spreading across major parts of the developing world, as will be seen in Chapter 3.

Conclusions

This chapter has outlined the major contours of the industrial grain-livestock complex in the temperate world, focusing on the USA, which has been so influential in the development of the global food economy through its tremendous productivity gains, exported surpluses, industrial innovations and the rise of its agro-TNCs. The structural changes and hyper-industrialization of agriculture have led to a great concentration of production and an extremely unequal productive base, trapping farmers in a cost–prize squeeze that carries pressures to grow in scale and hence mechanize in order to compete at lower and lower margins. The US industrial farm landscape is characterized by: the dislocation of its original inhabitants and the design of ahistorical, radically simplified and through-flow agro-ecosystems; the drawing down on 'natural accounts' in terms of soil and underground water; heavy external input usage centred on oil- and petroleum-based products and secured by the seed as commodity, including the rapid spread of GM crops; the cycling of increasing percentages of the cereal harvest through livestock; the warehousing of vast and growing populations of animals in unspeakably cruel conditions; fewer and ever larger farm operations; an insecure and 'super-exploited' workforce; and extraordinary per-farmer productivity. Similar tendencies are evident to varying degrees throughout the industrialized agro-exporting nations of the temperate world. For ordinary consumers in rich countries and the middle and upper classes in developing countries the result has been the rapid meatification of diets and a profusion of 'pseudo-variety', an incredible array of relatively cheap, standardized food that has been processed and recombined in a variety of ways and coupled with marketing drives to brand new consumer loyalties.

The large agro-subsidy regimes in the USA and the EU – which continued growing into the 1990s as agreements were taking shape to institutionally entrench the system – had a significant role in exacerbating the unevenness of agricultural landscapes while entrenching overproduction and an export imperative as the means of dealing with surpluses. This in turn led to the projection of distorted competitive pressures on to global markets, which has been particularly destabilizing for the 'unsubsidized Third World smallholders who cannot possibly compete with wildly under-priced cereals from abroad' (George 1990: 71), an issue that is examined in the following chapter. The forging of import dependencies in the 1960s and 1970s helped set the stage for explosive agro-TNC growth which further accelerated the convergence of diets across space. While the USA and Europe were at the forefront of the global expansion of the grain-livestock complex, highly industrialized surplus production from nations such as Brazil, Argentina, Canada, Australia, New Zealand and Thailand adds to the competitive dynamics in world markets with competition increasingly centred on Asia.

Agro-TNCs are the dominant actors shaping the temperate grain-livestock complex and with it the trajectory of the entire global food economy. As Heffernan et al. (1999: 2–3) put it, 'to understand the US food system, one must understand the global food system; to understand the global food system, one must understand the operations of the major global firms'. The rising control of agro-TNCs has had the revolutionary effect of despatializing food, with corporate-controlled circuits of accumulation increasingly governing how it is produced and distributed. Along with this increasing control over the input and output sides of agriculture, they have also sought to rewrite the rules of the system further in their favour, starting with national-level regulation and subsidy regimes and then setting their sights on a supranational constitution that would entrench and expand market access and sourcing flexibility, impede national supply management systems and price supports (which have been generally favoured by farmers) and force a US-styled intellectual property rights system on to the world.

These interests have had a major effect on the politics of multilateral agricultural trade regulation and in the design of the WTO, which is such an important part of the battle for the future of farming, as will be seen in Chapter 4.

Notes

1 The image of lands welded together borrows from Walt Whitman's imagery in 'Passage to India'.

2 Mason and Singer (1990) pioneered the study of modern factory farming practices, and there is a growing literature on the subject. In addition to their famous *Animal Factories*, the discussion in this section draws upon Nierenberg (2005), Midkiff (2004), Silverstein (1999), Fox (1999), and Davis (1996, 1995).

3 The discussion in this section, where not otherwise cited, draws especially from McMichael (2004b, 2000b), Friedmann (2004, 1994, 1993, 1990), Bernstein (2000), Berlan (1991) and Friedmann and McMichael (1989).

4 See the EWG's Farm Subsidy Database at <www.ewg.org/farm/> and the IATP's report on farm subsidies and below-cost feed crops at <www.agobservatory.org/library.cfm?refid=88122>.

5 The quotations in this and the following two paragraphs come from a large set of congressional testimonies given by USDA officials, corporate and large farm association lobbyists and elected representatives from concerned districts in order to prepare the US negotiating position in the lead-up to the WTO's Seattle Ministerial. The tetimonies made remarkably plain the vested interests at stake which were taken to the trade negotiations.

6 For more on the case in Schmeiser's words, see <www.percyschmeiser.com>.

THREE

From colonialism to global market integration in the South

Land and the colonial inheritance

European imperialism took dramatically different forms in the tropical and semi-tropical regions of the Americas, Asia and Africa than it did in the temperate neo-Europes. Extracting resource wealth with limited settlement generally pivoted on maintaining, exacerbating or establishing new inequalities in land and society, and binding the interests of local elites to the supply of a narrow range of agricultural and mineral commodity exports while undermining the local dynamics of innovation. The production of commodities such as sugar, cotton, coffee, tea, cocoa, palm oil, copra, jute and indigo in tropical and semi-tropical colonies was tied into Europe's industrialization and consumption patterns, with the combination of coffee or tea plus sugar connecting the exploitation of working people across continents in an especially intimate way – the products of slave, indentured or bonded labour and exploited peasantries in the colonies 'quelling hunger and numbing outrage' in the European working class toiling in the ghastly mines and factories of the age (Horowitz 1971: 1; Mintz 1985). In a basic sense, externally imbalanced trade relations had strong traction in the colonies through internally imbalanced social relations, though the system was ultimately also held together by violence or threat of force. Thus, the colonial inheritance for the developing world was not only about durable inequalities *between* nations but *within* them, and as Fanon (1982 [1963]: 44) put it, this was deeply embodied in the landscape: 'For a colonized people the most essential value, because the most concrete, is first and foremost the land: the land which will bring them bread and, above all, dignity.'

The human societies and agricultural landscapes of Latin America and the Caribbean were reconfigured the most profoundly. The violence of conquest, the exceptional indigenous vulnerability to 'Old World' diseases and the brutal labour regimes annihilated the indigenous population throughout most of the Caribbean in the space of one or two generations and exacted catastrophic declines on indigenous societies across much of Central and South America. As discussed in the previous chapter, the spread of alien livestock accompanied and even in places advanced ahead of European settlement in the Americas. Crosby (1972: 75) describes the tragic reconfiguration of the Caribbean: 'One who watched the Caribbean islands from outer space during the years from 1492–1550 or so might have surmised that the object of the game going on there was to replace the people with pigs, dogs, and cattle.' A similar process occurred on an even larger scale in other parts of the Americas. Present-day Mexico experienced the greatest population collapse in human history in the century after conquest as the human population fell from an estimated 25 million to 800,000 at the same time as the population of sheep was soaring and ravaging the landscape. An estimated 48 million cattle ranged across present-day Uruguay, Argentina and Paraguay by the mid-eighteenth century, having grown from only a few thousand that had been left behind by fleeing settlers less than a century earlier, and cattle also had a very significant role pushing into the interior of colonial Brazil (Cockburn 1996; Melville 1994; Crosby 1986, 1972).

Abetted by these animal pioneers, European colonialists seized the best agricultural land and established large plantations for lucrative exports like sugar and large locally oriented haciendas that supported the extraction of the region's great mineral wealth. European settlement was at first mostly along the coasts but expanded inland over time with surviving and rebounding indigenous populations 'exiled in their own land, condemned to an eternal exodus' as they were forced 'into the poorest areas – arid mountains, the middle of deserts – as the dominant civilization extended its frontiers' (Galeano 1973: 59). Latin American independence movements in the early nineteenth century did nothing to resolve the region's characteristic *latifundia–minifundia* (largeholder–smallholder) divide set in place by colonialism, nor did the so-called 'liberal reformers' in the late nineteenth century. While independence for most of Latin America largely entailed a shift of power in the colonial elite from the *peninsulares* (Spanish-born) to the *crillos* (Spanish descent, New World-born), a more hopeful and ultimately more tragic independence struggle occurred in Haiti, the site of the world's only successful slave revolution (1791–1804). After repelling the French and other European military interventions, key Haitian leaders were assassinated, the country was ostracized, saddled with an enormous debt by France and quickly fell into the hands of a small, comprador elite who maintained plantations and ruthlessly extracted surplus from the peasantry that emerged following the revolution (Trouillot 1994).

In Africa and Asia, prime agricultural land was drawn into export production geared towards European demand in different ways. The first basic approach involved coupling coercive labour regimes with unequal property rights, with the latter either established through dispossession or by manipulating existing imbalances. The second involved incorporating self-provisioning peasant farmers into market relations by making them wholly or partially dependent on petty commodity production through taxes or indebtedness. The social control needed to uphold the 'colonial division of labour' rested upon the minority of local elites in privileged commercial or administrative positions, a variety of unrepresentative political structures, the establishment of supportive legal and educational systems, the exacerbation or creation of ethnic divisions (divide and rule) and the repressive military apparatus. The net effect was not only the immediate pillage of biological and mineral resources and the exploitation of untold millions of people but the long-term rupturing of autonomous trajectories of economic, political and sociocultural development. With this rupture, as Rodney (1972) emphasized in the African context, the *locus of innovation* was eroded and shifted outside of the societies themselves, with Europe's early technological advantages most pronounced and decisive in weaponry and growing in scope and magnitude over time, from handicraft industries to agricultural practices, setting in train a key dynamic perpetuating uneven terms of trade long beyond colonialism.

But while the best agricultural lands were diverted by colonialism into export production in most of the tropical and semi-tropical world, establishing a deep-rooted and narrowly constituted commodity export dependence, this agro-export production was still restricted in geographic scope and food production retained a large degree of local orientation and self-sufficiency through the first half of the twentieth century. And many of these peasant economies did gain from at least one aspect of the European encounter – the introduction of crop varieties from other parts of the world that was coordinated through colonial botanical gardens; for instance, it is almost as hard to imagine many African agricultural systems today without maize as it is to imagine North American agriculture without wheat.

Struggles for (and against) land reform

Land was central to the course that decolonization would take, as Fanon insinuates in the quote above. At the end of the Second World War the great majority of the populations of decolonizing Asia, Africa and the Caribbean and neo-colonial Latin America still lived on small farms amid the land inequalities inherited from colonialism. Wrestling with these land inequalities meant class struggle as 'land reform necessarily implies a change in power relations in favour of those who physically work the land at the expense of those who traditionally accumulate the wealth derived from it' (Barraclough 1994: 17). This struggle, in turn, had clear international dimensions given the presence of foreign corporations entrenched in lucrative export enterprises, the international laws designed to protect externally controlled private property and the emerging geopolitics of the cold war, especially since the latent volatility of land inequalities had recently been brought into sharp focus by the revolutions in Mexico, Russia and China, and would become further evident in the coming decades in both victorious revolutions (e.g. Algeria, Cuba, Mozambique, Nicaragua, Vietnam, Zimbabwe) and sites of great rural violence and repression (e.g. Bolivia, Colombia, El Salvador, Guatemala) (Chaliand 1977; Wolf 1969). In this context, the USA saw the 'peasant question' as essentially being about how to demobilize the peasantries of the developing world and delink rural social movements from the urban anti-imperial, nationalist ones (Araghi 2000). Because of the inescapability of land to this question, land reform was 'practised along the entire spectrum of politics' and what it actually constituted varied greatly, ranging from redistribution into small private holdings, to reformist land tenure restructuring, to massive nationalization and collectivization schemes. As Hobsbawm (1994: 354–5) notes, 'between 1945 and 1950 almost half of the human race found themselves living in countries undergoing some kind of land reform'.

The largest of these was occurring in China where, prior to the revolution, more than half of the peasantry had been landless or semi-landless. Hinton (2000: 216), one of the great scholarly champions of China's land reforms, describes the assault on feudal ownership and obligations in 1947–52 as 'the most massive expropriation and distribution of property and repudiation of debt in world history'. The organization of peasants onto collectives or 'People's Communes' was at the core of the revolutionary project, and by the late 1950s private land ownership had been eliminated. Hinton (ibid., 1997) argues that China's land reform ultimately underpinned improved domestic food security for a rapidly growing population and laid a foundation for rural and industrial development, significantly expanding the reinvestment of agricultural surpluses, increasing demand for productive inputs and spurring local industries. But there is also a counter-interpretation of these reforms which points to the starvation of millions during the Great Famine of 1958–61 as damning evidence of their failure. This is a complicated debate and different scholars give dissimilar weight to factors such as: the loss of property and the new incentive system; the repression of the state; the abnormally bad weather of the period; the impact that the race to industrialize in the Great Leap Forward had on agriculture; and the legacy of stagnation and inequality left from the feudal order (i.e. would the famine have been much worse under the old order?).

Yet as disastrous as the Great Famine was and however it is interpreted, the resonance of the land issue in the Chinese revolution and the scale of China's land reform in the 1950s were very unsettling for the US government obsessed with containing the 'Red Menace', particularly considering that the Great Famine did not occur until the US policies for East Asia had already been established and that the world did not learn of the famine's magnitude until years later. China is widely seen to have inspired the USA to promote a different sort of land reform through its occupation policies and extensive post-war influence in Japan, South Korea and Taiwan: land reform in the name and defence of capitalism. Significant land redistributions were accomplished in these nations largely through state purchase and resale to the tillers, breaking down the supremacy of large rural landowners and creating countrysides of small- and medium-sized capitalist farmers. This served to demobilize more radical, anti-capitalist social currents and the dreaded 'communist subversion', as well as increasing agricultural output and establishing sizeable domestic markets for locally produced industrial goods (Huizer 1996). For this expansion of domestic markets these land reforms are often ascribed as a factor in the rapid industrialization that soon followed, together with a huge infusion of US development assistance and technology and US consent to the pursuit of highly protectionist policies towards infant industries – a unique combination rooted in these countries' particular geopolitical significance and proximity to China and the USSR.

While the USA also hoped to demobilize peasants as a potentially radical political force in Africa, Latin America, the Caribbean and the rest of Asia, the active support for land redistribution that had occurred in East Asia was not witnessed elsewhere. In Africa, beyond southern Africa and Kenya, the colonial ability to graft cash cropping on to peasant systems meant that land inequalities on the continent were not as extreme and hence land reform was less politically volatile. And the assassination of leaders such as Patrice Lumumba and Amilcar Cabral and the overthrow of Kwame Nkrumah in the era of decolonization dealt serious blows to the possibilities of transforming colonial dependencies, as did the repressive force of the apartheid regime in South Africa and later the great disappointment of Robert Mugabe following the Zimbabwean fight for independence. In Asia, the USA found anti-communist allies in tyrants like Ferdinand Marcos in the Philippines and General Suharto in Indonesia, who would successfully keep land questions bottled up for decades. Where peasant-led movements were seen to be too strong, and their direction was feared, the USA was prepared to intervene militarily, most disastrously in Indochina, where millions were killed.

In India, after China the next most populous country in the world and overwhelmingly poor and agrarian, a technical 'fix' was forged to partially mask ideological questions about land inequality, poverty, caste and food insecurity. Following independence, the assassination of Mahatma Gandhi and the partitioning of the subcontinent, India embarked on a race to modernize under the government of Jawaharlal Nehru (1947–64). Modest land reforms were initiated at the national level, and though some states pursued more substantive redistributions in many instances land reform initiatives largely served to widen the rural elite from a very small aristocracy to a new, somewhat larger – but still a decided minority – class of entrepreneurial commercial farmers. The limited nature of the land reforms of the 1950s and 1960s was a big part of the reason that grain production in India struggled to keep pace with surging population growth, and by the late 1960s stagnant grain production and rising food aid (India was one of the major destinations for US food aid at the time) was becoming worrying to the government.

But instead of deepening land reforms, which would come to prove a successful rural development and food security alternative in the states of Kerala and West Bengal, the primary response of the Indian government to national food insecurity problems was to turn to the technological package that came to be known as the Green Revolution, which was essentially the transfer of the industrial agriculture model from the temperate world: the introduction of new higher-yielding seed varieties (centred on adaptations of the big three cereal crops) coupled with fertilizers, agro-chemicals, irrigation (generally required in the absence of high rainfall conditions) and machinery (Shiva 1991). Because this provided 'a means of increasing food production without upsetting entrenched interests (as well as a means of providing increased revenues to Western firms supplying industrial inputs)', the Green Revolution effectively represented 'an alternative to agrarian reform' (George 1990: 184). In addition to abetting land consolidation by the larger entrepreneurial farmers who had greater access to these productivity enhancements, a further landscape inequality embedded in the Green Revolution has stemmed from the development of large-scale dams and irrigation projects which have typically excluded local people (Leonard and Manahan 2004). In India, the displacement of the rural poor by large-scale dam building has been put in sharpest focus by the long-running struggle over the Narmada Dam (Roy 1999). Yet while the Green Revolution had uneven social impacts which are discussed further below, it did bring large productivity gains in India as well as in the other developing countries where it was introduced (Conway 1997). The magnitude of these gains is suggested by the fact that while irrigated land comprises only about one-fifth of all farmland across the developing world it accounts for two-fifths of the developing world's total crop production and almost three-fifths of its grain production (FAO 2002a).

In Latin America and the Caribbean, the region with the world's most unequal distributions of land, the only major land reforms prior to the mid-twentieth century occurred after the Mexican Revolution (1910–17). The course and aftermath of the Mexican Revolution had a momentous impact on the character of agricultural development in Mexico, with implications for the rest of Latin America. Rallying cries of *'tierra y libertad'* (land and liberty) coupled with the leadership of Emiliano Zapata and his peasant army in the south carried the hope of a radical redistribution of property, and it is not hard to imagine what the success of a peasant-led revolution in one of Latin America's most influential nations could have meant for the region given its characteristic disparities, especially with significant peasant-based movements in Nicaragua, El Salvador, Colombia, Peru and Brazil also taking shape around this time. Following the revolution Mexico's new constitution legally enshrined a communal form of tenure, the *ejido* system, in which members maintain usage rights over an individual part of the community's land (the legal origins of which lie in the efforts of Spanish friars to set aside some land for indigenous subsistence cultivation in the face of colonial brutalities). The murder of Zapata in 1919, however, helped to derail radical reforms in the wake of the revolution and power quickly slid to the upper and middle classes in the Institutional Revolutionary Party (PRI). Only the Lázaro Cárdenas administration (1934–40) moved in Zapata's spirit, inscribing land reform into the constitution as an ongoing imperative to be pursued through the *ejido* system, and many peasants benefited from the 70 million hectares of land that were transferred from large estates into *ejidos*. But Cárdenas was followed by the degeneration of the PRI into a notoriously corrupt, patron-clientele state and momentum for land reform was overtaken by pressures from large- and medium-scale agro-capitalists, industrialists and commercial interests (Huizer 1996; Barry 1995). Though the *ejido* system continued to be viewed by peasants as their revolutionary entitlement, further pressure for land reform subsided and, as in India, the PRI sought to meet food security challenges for an exploding population with a technological fix as Mexico became a pioneering site of the Green Revolution.

Yet despite the social unrest and the volatility of inequitable land distributions throughout much of Latin America and the Caribbean, the USA adopted a watered-down advocacy of land reform compared to what it had done in Japan, South Korea and Taiwan. As had been done in East Asia, President Kennedy's 'Alliance for Progress' (1961) in the Americas framed land reform and the establishment of medium-sized private, family farms as a social and economic priority and a bulwark against rural social movements. But the action was not as aggressive as in East Asia. Though there was a near-doubling in the number of commercialized family farms in Latin America from 1950 to 1980 this was from a very small starting point, and in general the US-sanctioned land reforms in the region were modest in total size and focused on the colonization of marginal-quality state property and forest frontiers, while the *latifundia*, which dominated the best-quality land, remained virtually untouched (Araghi 2000; Barraclough 1994). The only way the *latifundia* were typically considered in such programmes was when governments would 'bail out large estate owners in economic difficulties by buying their lands for resettlement' (Barraclough 1994: 19).

The land inequalities in the region were made more galling and environmentally destructive by the tendency of large farms to under-utilize land and the increasing conversion of arable lands to cattle pastures for distant markets as many small farmers were displaced to serve the meat-intensive diets of 'the world's more affluent classes and societies' (Araghi 2000: 157). Central America and Amazonia provide the starkest examples, as extensive cattle ranching has been the leading cause of rainforest clearance despite the fact that the rangeland established there is among the most inefficient and erosion-prone in the world (the dense vegetation of the rainforests belies the thin soils underneath, as most nutrients are stored in the trees themselves, which rapidly reabsorb decomposing biomass; intensive erosion then occurs when canopies are cleared and these soils are exposed to the heavy rainfall of the tropics). In the 1980s, environmentalists drew attention to the so-called 'hamburger connection' linking deforestation, inequitable distributions of land and cattle ranging in Central America with demand from North American fast-food chains (Nations and Komer 1987; Myers 1981). Though the importance of external versus domestic demand has since been debated, the deforestation–inequality–ranching dynamics continue to be a major factor in the vulnerability of the region's remaining tropical forests.

In contrast to the minimalist land reforms that were sanctioned by the USA, the more substantive redistributions that brought real gains for the poor rural masses – most notably in Guatemala (1952– 54), Cuba (1959 and 1962), Peru (1958–74), Brazil (1962–64), Chile (1966–73) and Nicaragua (1979–86) – were 'in all cases undertaken in response to mass peasant mobilizations' or revolutionary pressures and triumphs (Veltmeyer 2005: 290). Not only did these have nothing to do with US pressure, they were instead met with its wrath. The USA had seen Latin American and the Caribbean as its imperial backyard since the Monroe Doctrine in the early nineteenth century and US capital had extensive, long-standing interests in various agro-export and mineral sectors which were woven together with the unbalanced social order throughout the region, most famously in Central America's 'banana republics'. To protect these interests or 'prevent the spread of communism', with the former often thinly veiled as the latter, the US government repeatedly intervened wherever serious land and social reforms were posed by revolutions or elected governments (e.g. in Bolivia, Brazil, Chile, Cuba, the Dominican Republic, El Salvador, Grenada, Guatemala, Jamaica, Nicaragua, Peru, Venezuela), using force, covert destabilization or both, routinely siding with brutal dictators and anti-democratic forces (Barraclough 1994; LaFaber 1983). The cases of Guatemala, Brazil, Chile, Nicaragua and Cuba are especially telling.

In Guatemala, the democratically elected government of Jacobo Arbenz (1951–54) initiated a reform of underused lands held by the US-owned United Fruit Company (beginning with an offer to purchase them), before being ousted by a US-backed and supervised military coup in 1954. The subsequent military regime reversed the reforms, persecuted participants and established death squads that would terrorize the rural poor in the decades of violence that followed, while holding in place a landscape in which 3 per cent of landowners controlled over 60 per cent of all arable land (Lovell 1995; Schlesinger and Kinzer 1982). The democratically elected government of João Goulart (1961–64) in Brazil pursued a relatively modest land reform which, similar to that of Arbenz in Guatemala, involved a confrontation with United Fruit over underused lands. Also like

Arbenz, Goulart was overthrown by a military coup in 1964 that received some covert CIA assistance and a congratulatory telegram from US president Johnson. The military junta promptly scrapped Goulart's land reform plans and established a programme to speed the colonization of the Amazon, in part to provide an outlet for the nation's rural poor and landless (Hecht and Cockburn 1989).

In Chile, where only 7 per cent of farms controlled 65 per cent of all arable land in the 1950s, land reforms were undertaken by the democratically elected governments of Eduardo Frei (1964–70) and Salvador Allende (1970–73). The Allende government took more aggressive aim at the *latifundia*, expropriating four thousand of the nation's largest farms and nationalizing 40 per cent of all arable land. These actions, along with the nationalization of the US-dominated copper mining industry, drew the ire of the local elite and the USA, and in 1973 Allende was overthrown in a military coup that again involved covert US support. The ensuing Pinochet dictatorship subsequently undid the previous decade's land reforms, smothered peasant organizations and abetted the reconcentration of land by a new class of agricultural entrepreneurs (Kay 2002).

But US backing could not sustain every dictator. In Nicaragua, the Sandinista revolution triumphed in 1979, toppling the Somoza dictatorship which ruled through two generations with US support while maintaining an iron grip on the country's land. For the Sandinistas, perceiving both the social injustice and ecological destruction of the landscape, land reform was at the heart of their revolutionary project and major redistributions were initiated. The Sandinistas' efforts were besieged and ultimately exhausted, however, by the US-supported paramilitaries (the Contras) loyal to the ousted dictatorship, before falling to electoral defeat (Faber 1993). Until Venezuela's recent democratic Bolivarian revolution, Cuba was the only revolution in the hemisphere to survive US pressure, and its experience with land reform is noteworthy. After ousting the Batista dictatorship, the revolutionary Cuban state quickly undertook a rigorous policy of expropriating all largeholdings, converting most large plantations and ranches into state farms in the first agrarian reform (1959) and taking control of more than three-fifths of all cultivated land in the second agrarian reform (1962). Most expropriated land was controlled by the state sector while some was given over to semi-independent cooperatives and a small percentage was kept as small farms. Because the USSR was paying grossly inflated prices to Cuba for its sugar, however, (by the 1980s more than five times above world market levels), the state had little incentive to diversify. Instead, the sugar industry was reorganized and production increased through mechanization and the surpluses used to subsidize Cuba's social achievements in health and education. While the failings of this model – in particular, the fragility of food security, the oil dependence of industrialized agriculture and the alienation of workers on collective farms – were exposed by its collapse after the demise of the Soviet empire, it is nevertheless instructive to see what was accomplished when agro-export earnings were used for social ends rather than monopolized by a small landed elite or by foreign companies. Cuba's second experience with land reform from 1990 onwards also has very important lessons which are discussed in Chapter 5 (Funes et al. 2002; Rosset 2000).

In short, while land was central to many anti-colonial struggles and land reform was occurring widely in various forms in the 1950s and 1960s, there was also intensive, often decisive opposition from landed elites, transnational capital and military force which bled the momentum for land reform, in many instances all too literally. After the 1960s there were few serious land reform initiatives or successes. By the late 1990s in Latin America three-quarters of all farmers had access to only 10 per cent of all arable land (Veltmeyer 2005), and 'the striking social inequality in Latin America can hardly be unconnected with the equally striking absence of systematic agrarian reform from so many of its countries' (Hobsbawm 1994: 356). The same is true for southern Africa and Kenya, the Caribbean outside of Cuba and large parts of Asia. The suppression of land reform thus played a major role in the course of agricultural and more broadly economic and social development in much of the developing world and added fuel to the dynamics of both rapid urbanization and environmental degradation (Berry 1997; Plant 1993; Thiesenhusen 1991). Yet as the neoliberal development model was becoming dominant in the 1980s through the application of structural adjustment programmes (SAPs), the problems posed by the issue of land distribution were largely severed from their history and shrouded in highly abstracted logic (discussed later in this chapter).

The marginals: the construction of food import dependencies

Colonialism restructured prime agricultural landscapes to produce commodities for export throughout much of the tropics and semi-tropics as well as altering, to varying degrees, the social relations through which food was produced and exchanged. In some instances the importation of grains (mainly wheat) and dried fish and meats had also been introduced. But this importation was generally at small levels and, with the bounds of time and space still significant, agricultural production was necessarily largely oriented towards local consumption. Prior to the Second World War only a small range of key food items were transported in significant volumes over long distances, and colonies by and large produced their own food supply and maintained an agricultural trade surplus (while holding major deficits in manufactured goods). This is encapsulated in a contemporary British colonial report which noted that:

> ... in almost all colonies cash crops have been developed to a considerable extent, and every day money comes to play a more and more important part ... [But] in most parts of the Colonial Empire the bulk of foodstuffs consumed are produced within the territory of consumption ... [and] imports form a negligible portion of total consumption. (EACCNCE 1939: 42)

As discussed in Chapter 2, international agricultural trade began to change rapidly after the Second World War, driven by the structural surpluses of the temperate breadbaskets that were seeking new outlets. Irrespective of whether they read US and European food aid and subsidized surplus dumping as being motivated by domestic structural pressures, the desire to establish new export markets, support for cold war allegiance or naïve benevolence, governments in many newly independent developing countries were only too eager to receive the cheap food. With dreams of modernity embedded in plans for rapid industrialization and urbanization seen as a necessary and inevitable social transformation, cheap food imports were framed as a means to speed the development of an industrial working class while containing wage pressures (given the importance of food in the cost of living and hence in the price of labour). This vision also somewhat ironically combined cheap food imports with the widely accepted goal of import-substitution industrialization, which was pursued at different points along the ideological spectrum.[1]

For African governments facing rapidly growing populations but denied the technological fix of the Green Revolution by climatic factors, cheap grain imports were taken to be even more necessary in the push to modernize. Nigeria was a classic case, as government plans sought and failed to establish industrialized wheat production and moved the country from being a non-existent grain importer at its independence in 1960 to being a large cereal importer in the space of two decades (Andrea and Beckman 1985). Coupled with these imports and in line with the objectives of urbanization and industrialization, state-run marketing boards also commonly served to contain the prices of domestically produced food, meaning that poor rural producers were in effect subsidizing relatively wealthier urban consumers – a key reason why modernization-oriented development planning was later described as having an 'urban bias' (Lipton 1977). In addition to the theorized benefits for urbanization and industrialization, in practice food aid and cheap food policies were sometimes manipulated by political elites to solidify their authority through patronage-motivated handouts (Leys 1996; George 1990).

Cheap food imports were also seen to allow increasing agro-export production, typically in the rut carved by colonialism, with the hope that the increased foreign exchange earnings and their taxation could be used to fuel broader national development goals, especially import-substitution industrialization and infrastructure development. The drive to expand agro-export production took different forms, including increased technology on plantations, the expansion of plantation frontiers and the incorporation of ever more peasants into market relations (Bernstein 2000). Increasingly capitalized agricultural techniques were propelled by what Bernstein (1990: 74) describes as 'unholy alliances of state, foreign aid, and international capital', though Soviet-influenced socialist states also shared the belief that large-scale industrialized agriculture carried efficiency gains.

The promotion of urban-biased cheap food policies together with the intensification of agro-exporting and the suppression of land reform resulted in rising import dependence, as many domestically oriented small farmers were negatively affected by the temperate grains that flooded markets, depressed food prices and changed tastes (e.g. wheat and flour replacing coarse grains). The net result was that many agrarian societies quickly became far more dependent on agricultural trade shortly after independence than they had been under colonial rule. Within a matter of decades developing countries collectively 'rose from being practically non-existent as [grain] importers to taking almost half of world imports in 1971 – and at their peak in 1978, they bought 78 per cent of US wheat exports', which was a critical part of the deepening integration into the global food economy (Friedmann 1990: 20, 1993). When food imports were cheap and tropical commodity prices relatively buoyant in the 1950s and 1960s, being a price-taker in commodity markets seemed illusorily viable and agro-export growth helped finance development projects. But for those parts of the developing world that had become dependent on imported food staples through food aid and concessionary sales, and which had failed to significantly industrialize or were not endowed with oil – a list that included most of the world's LDCs – this became an increasingly perilous route to food security after

1972–73 as US food aid transmuted into commercial exports at a time of temporary global market shortages and the oil price shock sent global capitalism into a recessionary spiral.

Most developing countries were hit hard by this crisis in global capitalism as nearly all traditional commodities, both agricultural and non-agricultural, began a steady descent on global markets vis-à-vis the much more diverse basket of manufactured goods that were being imported (Robbins 2003; FAO 2003; Khor 2002). These declining terms of trade for tropical commodities were rooted in a combination of factors. The rising costs of oil and energy in the 1970s made the relative value of manufactured goods rise above that of unrefined commodities. Differential prices also reflected the fact that organized labour in the industrialized West was able to achieve a rising social wage (i.e. pushing up wages and benefits ahead of inflation as part of its Keynesian cold war pact with the state–capital nexus), while the share of the gross product held by working people in developing countries generally hovered much closer to a 'physiological wage' (i.e. the level that allows for the reproduction of labour). The fact that trade in most tropical commodities is significantly monopolized by a small number of massive, vertically integrated agro-food TNCs has also greatly affected country earnings as corporations can manipulate prices and reduce export tax bills through intra-company transactions or what is known as transfer pricing strategies.

Another important factor in the declining terms of trade for tropical commodities is the matter of structural overproduction, especially as some new producing areas were opened – recalling how even small surpluses in markets can place large downward pressure on prices, which was an important part of why the export imperative was built into temperate agricultures. Compounding this, some key export markets for tropical agricultural commodities were shrinking as a result of their susceptibility to industrial substitution, among these cane sugar (high-fructose corn syrup, artificial sweeteners and beet sugar), cotton (synthetic fibres) and palm oil (soy and canola oils). And because production is typically spread out across a number of countries it is difficult to control with the net result being a spiral in which each would generally seek to make up for falling prices with expanded production and hence further entrench these dynamics (Robbins 2003). The imbalances associated with declining terms of trade have also been accentuated by the selective protectionism of industrialized countries against processed, semi-processed and directly competitive goods from the developing world (i.e. placing higher tariffs on processed than unprocessed commodity forms) with these discriminatory 'tariff peaks' making it difficult to move out of the role of raw material supplier and limiting the share earned by developing countries in commodity chains.

While individual nations were unable to affect commodity prices received in world markets, collectively some gains were made. The newly independent nations of the Africa, Caribbean and Pacific (ACP) group were shielded from the full brunt of these falling terms of trade by protected markets in the EC which had been established through the Lomé Convention in 1975. Owing to the sense of responsibility Britain and France felt for their former colonies, the central tenets of Lomé were non-reciprocal, duty-free access and stabilized prices for all industrial and most agricultural exports from the ACP to the EC, and increased development aid and industrial cooperation for ACP countries. Sugar is one notable commodity whose prices were long insulated from world market fluctuations and declines by Lomé. Yet while Lomé was a modest effort to reduce the vulnerability of narrowly structured economies to price fluctuations, in stabilizing commodity export earnings it also threatened to perpetuate these very dependencies and with them often exploitative systems of production (e.g. sugar plantations). As such, it was essentially a palliative rather than a cure for the uneven system of commodity dependence forged by colonialism.

Eventually the rising dependence on food aid and concessionary sales coupled with the goal of expanding traditional agro-exports came into focus as having been a wolf in sheep's clothing for many of the world's poorest countries. As Friedmann (1993: 38) puts it, 'by the early 1970s ... the food regime had caught the Third World in a scissors. One blade was food import dependency. The other blade was declining revenues from traditional exports of tropical crops.' Where peasantries were transformed into petty commodity producers dependent on the purchase of food staples and costly inputs, vulnerability to hunger and famine tended to increase. With production disembedded from local ecological conditions and mediated by new technologies (e.g. the purchase of seeds and inputs) and exchange severed from old social relations and tied into market pressures, environmental risks such as drought became more destabilizing, nowhere more acutely than in sub-Saharan Africa (Watts 1983). A particularly devastating case of how food import dependence and the privileging of commodity exports led to declining food security occurred in the Sahelian region of Africa during the early 1980s. As drought and hunger were gripping the Sahel in 1983–84, five nations in the region – Burkina Faso, Chad, Mali, Niger and Senegal – imported a record amount of cereals while exporting record volumes of cotton as its price was falling on world markets.

The mounting and painful evidence of global economic imbalances drove many developing countries towards increased assertiveness and organization in the 1970s, including calls for a New International Economic Order (NIEO) and formations like the Group of 77 (G-77) and the Non-Aligned Movement (NAM). Within the United Nations system, the NIEO movement had a significant articulation in the United Nations Conference on Trade and Development (UNCTAD), which became a valuable storehouse of data on commodity production and pricing patterns and declining terms of trade and pressed for efforts to stabilize commodity prices such as through supply management. As will be seen later in this chapter, however, these efforts faced a backlash in the 1980s from which they have not yet recovered, and terms of trade for tropical commodities continued worsening. For the world's LDCs, the terms of trade decreased by roughly *half* in the quarter-century following 1973 (UNDP 1998), what Robbins (2003) aptly calls 'the tropical commodities disaster'. In the face of these structural imbalances, the LDCs have been at the farthest margins of the multilateral negotiations giving shape to the institutions that govern the global trading system, as will be seen in Chapter 4.

The powerhouses i: prioritizing domestic food self-sufficiency

The contemporary agrarian question is posed in its grandest absolute dimensions in China and India. Though fast-growing industrial and high-tech giants, with China oft described as the new workshop of the world, both nations remain heavily rural and are together home to more than half of the world's agrarian population: roughly 70 per cent of China's 1.3 billion people (900 million) and more than three-quarters of India's 1.1 billion people (770 million) live in rural areas, and the majority of these still depend on agriculture for their living. From nationalist, geostrategic and economic motivations, the governments in both countries have sought to avoid food import dependence, closely controlling trade and pursuing very different agrarian reforms aimed at enhancing food self-sufficiency. Because of this, China and India have a small place in global agricultural trade relative to their massive populations and booming economies, and they are therefore seen by agro-food TNCs as having the potential for vast market growth if trade could be liberalized. For both, agricultural trade liberalization would exact an enormous social cost.

While the effectiveness of China's first land reform continues to be debated, the inescapable backdrop for this debate must be the dramatic ascent of China's average life expectancy from around forty years at the time of the revolution to nearly seventy years by the late 1970s during a period of rapid population growth, which would seem to provide some solid basic evidence for Hinton's thesis about the effectiveness of China's first land reform. The model did not last long beyond Mao's death in 1976, however. Much as agriculture had been central to Mao's communist vision so too was it central to China's transition to state-managed capitalism, as the People's Communes were soon after dismantled in a second land reform that allocated individual pieces of land to small farmers and allowed them to farm independently (though not given outright ownership). This was accompanied by newly loosened profit incentives for farmers and state support for high-input, high-yield farming through the expansion of irrigation systems and subsidies for enhanced seeds and fertilizers, and per capita food production grew significantly thereafter (Sen 1989).

In contrast to the intense debate that has surrounded China's land reforms, very little attention has been given to understanding how they have been interwoven with another enormous agricultural transformation: the incredibly rapid meatification of Chinese diets, which is also fast being embedded in China's warp-speed industrialization with increasing factory-farm production. From 1961 to 2004 the volume of meat production in China grew by an astonishing factor of 29, and while its share of the world population has remained relatively constant at 22 per cent over the past half-century its share of the world's meat production has risen dramatically from 4 per cent in 1961, to 11 per cent in 1980, to 29 per cent in 2004, including *half* of all the pigs produced on earth today (FAOSTAT). The per capita consumption of both meat and milk has roughly doubled since the reforms of the early 1980s, or roughly a single generation. In describing the growth in meat consumption in China in the 1990s, *The Economist* (2003: 4) noted that 'what took a century to happen in the West has taken a decade in China', while pointing out some important class and cultural dimensions of this change with the affluent at the vanguard of substituting meat for grains. Every year roughly two hundred new KFCs and one hundred new McDonald's are opening in China, with US fast-food giant Yum! Brands (the KFC parent) describing its goal as: 'to be the leader in every significant food service category in mainland China'.[2] The ambitions of agro-food TNCs in the face of China's economic boom and dietary revolution are beautifully summed up by a top Nestlé executive in China: 'This is the fastest and most competitive market in the world ... Growth in China is up to your imagination' (quoted in ibid.: 8).

Already 15 per cent of China's pig and chicken meat is produced on factory farms, though these methods have only recently begun increasing in frequency there, and this is poised to soar as factory farming is at the centre of government plans to double the total value of animal production within a decade (Nierenberg 2005). This policy runs entirely counter to the recommendations from long-term dietary and epidemiological research in China, which advocates the health benefits of less meat-intensive diets (Campbell and Campbell 2005; Chen et al. 1990), and it is notable that Chinese obesity rates are now rising quickly. From negligible levels a generation ago an unprecedented percentage of the Chinese population is now considered to be overweight (15 per cent), with roughly 90 million obese according to the WHO, rates that have grown even faster than they did in the USA between 1964 and 1990 when obesity was becoming rampant there (Economist 2003).

China maintained a net cereal surplus into the 1990s but the increasing cycling of grain through livestock has drawn it into a cereal deficit, and its current course will have major implications for the future of the global food economy, particularly in terms of rising demand for maize and soy feed. Enormous domestic questions are also posed as China's labour-intensive agricultural system becomes increasingly capitalized. Urbanization will have to absorb hundreds of millions of rural Chinese, with some estimates suggesting as many as half a billion in the coming decade 'if rural mayhem and revolt is to be avoided' (Harvey 2003: 161). A recent report noted that 'violent protests are convulsing the Chinese countryside with ever greater frequency', citing a large rise in government-documented 'public order disturbances' from 10,000 in 1990 to 87,000 in 2005, along with the fact that 'income disparity between the urban rich and the rural poor is at its widest since the People's Republic was founded in 1949', and suggesting that the government now faces the challenge of having to 'assuage rural discontent before it hardens into a wider, more flammable agrarian revolt' (Time 2006: 22, 24).

Like China, India today faces a similarly worrying rural crisis teetering near something 'more flammable', but one with very different roots. As noted, India was the contested, archetypal symbol of agricultural modernization in the developing world, embodying both the promises and the polarity of the Green Revolution – its advocates describing it as a 'spectacular success' story (Chrispeels 2000: 3; Conway 1997) and its critics railing against it as class and gendered 'violence' (Shiva 1993, 1991). The enhanced seeds of the Green Revolution were developed in the 1950s in the research conducted by US scientist Norman Borlaug and his associates (endowed by the Rockefeller Foundation, a US philanthropic organization), which brought rising wheat and maize yields on irrigated lands in Mexico. The enhanced seed–input package innovations achieved in maize and wheat were also extended to rice research at the International Rice Research Institute in the Philippines. In each case the focus was on adaptations suited to high-quality, irrigable or high-rainfall lands, monoculture planting and machinery, a research agenda that brought impressive yields gains in some places but ignored the needs of poorer farmers situated on less endowed lands. Together these packages were disseminated through scientists in the system of International Agricultural Resource Centres, a quasi-public domain (part of the Consultative Group on International Agricultural Research, based at the World Bank), which meant that the new seed varieties were distributed to farmers without intellectual property rights licensing protections, and they were also often coupled with state subsidies for fertilizer, pesticides and irrigation. Yet though the work was ostensibly done in the name of international development, modernization and saving the world from hunger, for which Borlaug was awarded the Nobel Prize in 1970, US seed and agro-chemical companies were keenly attuned to the possibilities for accumulation these innovations prised open (Kloppenburg 2004).

Agricultural production increased significantly not long after the introduction of the Green Revolution to India, and though the productivity gains were spatially concentrated in the north-west cereal heartlands of the Punjab, Haryana and western Uttar Pradesh, on an aggregate national level they soon began to outpace the rapid population growth that was occurring. India attained grain self-sufficiency in a relatively short period of time, which allowed the government to replace its receipt of food aid with a state-controlled food grains buffer stock to be amassed and dispersed as necessary with the goal of maintaining prices and national food security for when adverse monsoon conditions brought poor harvests and shortages. But because the food sufficiency objective was defined as ensuring that aggregate demand could be met through domestically produced grains and supplied at relatively consistent prices in markets – and not that every household was food secure – great distributive inequalities within India meant that widespread, persistent undernourishment continued in spite of an effectively organized national surplus grain stock (Swaminathan 2006).

The now well-known criticisms of the Green Revolution on the consumption side are that the food security miracle was seriously overstated given the lack of attention to distribution and on the production side that the costly (though state-subsidized) input package meant that the reforms tended to reinforce existing privilege or open new possibilities for land accumulation. Amid the productivity gains Green Revolution proponents 'rarely bothered to ask "Production by whom? and for whom?"' (George 1990: 184); or what impacts a capital-intensive agricultural model would have on labour-intensive agricultural systems and employment-scarce settings; or what the long-term ecological impacts would be as control over the agricultural systems shifted from local communities to distant research labs and, in time, to profit-seeking agro-input TNCs. Instead, any such social or environmental questions were simply subsumed under 'the ideology of inevitable technological progress' (Middendorf et al. 2000: 117). This also related to the massive dam-building projects which, as noted, have had a very uneven distribution of benefits and costs.

Larger farmers with capital or access to credit and better-endowed lands could afford the necessary inputs and irrigation costs and success with the new technologies encouraged expansion, allowing them to dispossess smaller farmers who were excluded from these technologies by lack of scale and capital, and who struggled to compete against the price pressures caused by productivity gains – with the capitalized entrepreneurial farmers a key new force in the social polarity of rural areas. The exacerbation of social inequalities between large landholders, small farmers and the landless in Green Revolution landscapes also had marked gendered dimensions as many women farmers lacked adequate tenure, legal rights and representation and voice within households and communities and fared disproportionately poorly in gaining access to new technologies and in competitive land markets. And as many women farmers were displaced and transformed into part-time workers their skills, knowledge of seeds and ultimately wage levels were devalued by mechanization and standardization (Shiva 1999, 1991; Agarwal 1994).

Ecologically, the Green Revolution dramatically increased the input intensity, petroleum dependence (in fuels and inputs) and bio-simplification of agriculture. The area of pesticide-sprayed land in the developing world grew more than thirteen-fold between 1960 and 1980 with India at the forefront (Conway 1997), and India's estimated annual synthetic fertilizer use grew from 0.1 million tons in the early 1950s to 3.4 million tons by the mid-1970s (GRAIN and RAFI 1996). This input treadmill carries the standard concerns about soil and water pollution, and the particular vulnerability of monocultures to the erosional and leaching forces of heavy rainfall is obviously of great concern to the long-term productivity of soils in monsoon areas. Just like the process described in the previous chapter, the growth of industrial monocultures transformed farmers into recipients rather than agents of on-farm research and innovation, and drove them into the market for hitherto common property or on-farm generated resources with the commodification of seeds and the rising use of external inputs (Kloppenburg 2004; Shiva 1993, 1991). Thus, the Green Revolution sowed the seeds, figuratively and materially, of growing corporate control over the agricultural system and the dramatic reduction of crop diversity. Indian farmers grew as many as 50,000 different strains of rice prior to the Green Revolution, but by the 1990s this had been reduced to a mere dozen, reflecting a logic that does not value the *in situ* conservation of diverse crop varieties in diverse agro-ecosystems but rather sees only the need for 'high-tech' *ex situ* conservation in gene banks and laboratories (GRAIN and RAFI 1996).

The productivity gains of the Green Revolution were largely exhausted by the 1990s, at which point India's population growth began to increase faster than food grain production for the first time since the 1960s, while investment in agriculture was declining and the undernourished still counted in the hundreds of millions. Meanwhile, with its high-tech sector leading national economic growth, India now possesses arguably the world's largest middle class by global consumption levels and, much as in China, the growing upper and middle classes are leading the shift towards more livestock-intensive diets, though not at the same pace. Though there are religious and cultural reasons why per capita livestock production is relatively lower in India than in other industrializing Asian economies, it is nevertheless rising quickly, especially in dairy, poultry meat and eggs, with the result that rising volumes of grain are now being channelled through livestock. By volume, India is now the world's leader in aggregate dairy production and the world's fifth-largest producer of broiler chickens and eggs, the latter partly through the recent rise in the factory farming of chickens (Nierenberg 2005, 2003; Gandhi 1999).

Given the extent of persistent hunger and the growing grain demands of increased factory farming, some have advocated that another technological fix is needed, and the Indian government recently began allowing some GMO crop trials in the face of great resistance (Ramakrishnan 2006) and has faced consistent pressure for trade liberalization at multilateral negotiations. At the same time, signs of agrarian distress are rampant, disturbingly marked by the soaring number of farmer suicides. Suri (2006: 1523) notes that while 'agrarian distress is not new to India … farmers' suicides are', insisting that 'what is happening today seems to be qualitatively different', and India's federal agriculture minister recently conceded to parliament that more than 100,000 farmers committed suicide between 1993 and 2003 with debt identified as the most important factor (Indian Express 2006). India's Ministry of Agriculture also recently acknowledged the broad rural crisis in a policy document:

> … agriculture has become a relatively unrewarding profession due to a generally unfavourable price regime and low value addition, causing abandoning of farming and increasing migration from the rural areas. The situation is likely to be exacerbated further in the wake of integration of agricultural trade in the global system, unless immediate corrective measures are taken. (Quoted in Suri 2006: 1523; see also Frontline 2006)

In addition to trade liberalization, the prospects of a second technological-agrarian revolution would only make these social pressures more daunting.

In light of their contemporary agrarian questions and the pressures they face to liberalize agricultural trade, both China and India would do well to observe the experience of Mexico, which is sometimes described as a test subject for the developing world as regards the outcomes of multilateral trade discipline. Further, like China and India, Mexico experienced rapid population growth along with uneven industrialization and urbanization in the twentieth century, and undertook an agrarian transformation with elements of both the Chinese and Indian models. On one hand, the *ejido* system that followed the Mexican Revolution protected a measure of small-farmer and indigenous land and food security in perpetuity, and with it the centrality of maize production, which had been at the heart of food security and cultural identity for millennia. On the other hand, Mexico was the birthplace of the Green Revolution, and large industrialized farms achieved major yield and productivity gains. This dualistic landscape of communally controlled small-farm holdings and industrialized, high-yield production together made Mexico largely food self-sufficient into the 1980s, even as its population grew more than sevenfold over the course of the twentieth century.

The onset of structural adjustment in the 1980s, however, brought a dose of trade liberalization to agriculture, cutting import tariffs and quotas, along with policy changes favouring export growth. And as some high-value Mexican exports such as fresh fruit and vegetables gained a foothold in off-season North American markets (part of the dramatic rise in the all-season availability of fresh fruits and vegetables there) commercialized farmers experienced increasing incentives 'not to produce the necessary but rather the profitable' (Barkin 1987: 287). In tracing the tomato commodity chain, Barndt (2002) provides a detailed insight into some of the social costs underpinning this growth. The flipside of the rising export orientation of these successful farms was that cheap grain imports began rising (Barry 1995; Barkin 1987), and these became a torrent following the onset of NAFTA, which was signed between the USA, Mexico and Canada in 1994. NAFTA affected Mexican agriculture in two fundamental ways. First, state support programmes for Mexican agriculture (e.g. government-subsidized credit and marketing boards) were cut and the state's formal (albeit long inert) obligation to land reform was terminated, as a precondition of Mexico's admission to NAFTA was its 1992 constitutional amendment to allow the privatization of the communal *ejidos* – which had previously been protected from sale or fragmentation. Second, NAFTA entrenched the market liberalization begun under adjustment, enhancing openness to imports while reducing domestic price supports for farmers. Understanding these changes as a further assault on both indigenous livelihoods and culture, the Zapatistas chose the onset of NAFTA to mark their uprising (see Chapter 5).

In the decade following NAFTA, Mexico's total agro-imports increased by 50 per cent, led by the more than tripling of maize imports from the USA, its modest agricultural trade surplus became a sizeable deficit, maize prices collapsed amid the flood of cheap, highly subsidized US imports (which now account for roughly one-third of national maize consumption) and more than 1 million people were squeezed out of agriculture (Rosset 2006). At a rally in Mexico City in 2003 small farmers angrily exclaimed that 'we are the people of maize, and now we are importing it from the north!'[3] Meanwhile, at the same time as subsidized US maize is increasingly squeezing small farmers in Mexico and into neighbouring Central America, one of the largest net sources of foreign exchange in the region is the remitted earnings from long- and short-term Mexican and Central American migrants in the USA, where they comprise a critical source of cheap labour in the farm economy – a conspicuous feedback loop in a very contradictory food economy.

From colonialism to global market integration

The powerhouses ii: building competitive agro-export platforms

While agrarian change in some large developing countries focused primarily on domestic food security, a few relatively wealthy states in Latin America and Asia had a different goal: to enhance their agro-export earnings by rapidly industrializing and diversifying production, spurred by both government investments and policy incentives to agro-TNCs or domestic capital. This included the expansion of grain, soy and livestock production (especially in the southern cone of South America) and the mechanization of 'traditional' exports like sugar, as well as shifts into new or 'non-traditional' agro-exports such as fresh fruit and vegetables, high-value-added processing, horticulture and aquaculture and shellfish (with the latter revolutionizing the coastal ecology in large parts of South and South-East Asia), production which came to occupy an important position in the global food economy in a relatively short space of time (Friedmann 1993). Unfortunately, this agro-industrialization and export growth has all too often been built on the denial or reversal of land reforms or the expansion of large, commercialized farmers within competitive land markets. Further, the everyday toxicity facing estate workers on high-input modern farms poses risks of short-term poisonings (WHO 1992; WHO and UNDP 1990) as well as long-term cumulative health implications (Mendis and Van Bers 1999; Murray 1995). In recent years, a series of lawsuits filed by Central American estate workers made sterile by their exposure to chemicals has drawn some international attention to these chronic health problems, as well as to the difficulty of poor, unorganized workers finding justice when pitted against large plantations and chemical companies.

In much the same way as India is both the archetype and the biggest case of the Green Revolution, Brazil is an especially illustrative and influential case of the growth, inequality and ecological costs of the rapidly industrialized agro-export model in the South since the 1970s. In 1970, Brazil was relatively insignificant in global agricultural trade apart from its old position as a leading sugar and coffee exporter. Today it is also one of the world's biggest-volume and most competitive exporters in the grain, soy and livestock production that lie at the heart of the global food economy, with the USDA recently identifying Brazil as a major competitor to US exports (Nierenberg 2005). Along with Brazil's rapid agro-export growth has come the growing presence of many of the world's largest agro-TNCs, including agro-food TNCs like Bunge, ADM, Nestlé and Danone and agro-input TNCs like Monsanto, BASF, Bayer, Syngenta and Novartis

(MST Secretariat 2006). This agro-export growth has occurred in an extremely uneven landscape; despite 'a rural exodus of staggering proportions (30 million over the past 25 years)', there are still nearly 5 million rural families in Brazil who are landless, while only 3 per cent of the population controls roughly two-thirds of all arable land (Veltmeyer 2005: 290; Araghi 2000). This inequality gave rise to the MST (Movimento dos Trabalhadores Rurais Sem Terra, or Landless Workers Movement), which has grown into one of the world's most prominent rural social movements and a major political force in Brazil, pushing for radical change through direct action (e.g. land occupations, transforming non-productive estates) and different forms of political education and advocacy (Branford and Rocha 2002).

In addition to its social costs, Brazil's agro-export growth has also exacted a colossal toll on the environment. The destruction of the Brazilian Amazon accelerated in the 1960s as the military junta targeted it as a frontier to be colonized, a potential cattle range for exporting and an internal outlet for its agrarian question in the uneven and more densely settled landscapes in the south and east. To this end, land, subsidies and tax concessions were provided and a massive road-building scheme was initiated. While some small farmers and landless people moved west to modest plots, the process of colonization was dominated by the Brazilian elite, which acquired and cleared vast tracts of land and stocked it with cattle, reaping subsidies and tax breaks and hoping for speculative profits (Hecht and Cockburn 1989). As Cockburn (1996: 36) summarizes it, 'big ranchers, rather than the peasant settler-pyromaniacs of song and story, accounted for most of the destruction'. For decades, most Amazonian cattle were consumed in Brazil's industrial south, but as Brazil's beef exports rose in the 1990s a new export dimension was added to Amazonian ranching. Between 1995 and 2003 alone Brazil's beef exports tripled to a value of US$1.5 billion (Nierenberg 2005; Vidal 2004). In addition to the continuing razing of the Amazon for cattle ranching – surely the world's most ecologically costly pasture land – Brazil's frontier forests are also being cleared for another key element of the grain-livestock complex: soybeans. Brazil's soy boom is feeding a rapid escalation in factory farming which has made it one of the world's largest and fastest-growing producers and exporters of poultry and pig meat (Nierenberg 2005). It is also helping to propel the growth of factory farms in other countries that are short of domestic feedstock.

Perhaps the plainest evidence of the contradictions of Brazil's agro-industrial model is the fact that extensive malnutrition and hunger persisted amid its large and growing agricultural trade surpluses, and most pervasively so in rural areas. In spite of Brazil's agricultural and industrial growth, at the time the populist president Luiz Inácio Lula da Silva and the Workers' Party came to power in 2002 nearly a quarter of Brazil's population (roughly 40 of 170 million people) were living on less than US$1 a day (FAO 2003). This prompted Lula to pledge to eradicate hunger within his four-year term and establish the Zero Hunger Project in 2002, which couples a range of short-term ameliorative programme initiatives (e.g. household food budgetary support for families, school and workplace feeding programmes, employment training programmes linked to food assistance) with potentially more substantive programmes aimed at connecting the demand stimulus of this public food safety net to production support for small- and medium-scale farmers (e.g. credit, insurance, extension services); in short, aiming to connect poor rural and urban consumers to small producers in sustainable networks (FAO 2003). Yet while Lula's initial victory and second-term triumph in 2006 were based in large measure on the aspirations of Brazil's poorer classes, they have nevertheless been a great disappointment to the MST for failing to live up to promises on land reform, treading very gently on the real fulcrum of food insecurity and rural poverty in the world's most inequitable landscape – access to land – while initiating more moderate food access initiatives, maintaining the preceding focus on export expansion and accommodating the interests of the large commercialized farmers, domestic agribusiness and agro-TNCs (Stedile 2007; MST Secretariat 2006).

As in Brazil, the emergence of competitive agro-export platforms in the other powerhouses of the developing world has tended to reproduce or deepen the unevenness of agricultural landscapes, in contrast to the oft-repeated promise made by free trade advocates that export growth is the means to enhancing agricultural development. One common way in which the developmental (or 'pro-poor') case for increased agro-exporting has been made is in the promotion of 'non-traditional' crops, particularly fruit and vegetables, which generally face lower trade barriers than do processed goods in rich-country markets (provided they are not in direct competition in terms of variety or season) (Thrupp et al. 1995). Part of this case rests on the relatively higher per-volume prices that can make smaller land sizes more profitable than for traditional tropical commodities, especially in the context of the tropical commodities disaster. Where non-traditional agro-exports have successfully been established, however, a number of barriers have made it difficult for small farmers to establish production and compete in new marketing networks, such as: poor infrastructure in remote or hilly areas; large initial investment demands (e.g. fruit tree seedlings, irrigation and other inputs, with fruit trees facing the prospect of no early return); limited capital or access to credit; rigorous marketing requirements and costly storage facilities stemming from high perishability and cosmetic demands; and the preference of exporters to buy from larger, well-endowed farmers and avoid the high transaction costs of dealing with many smaller ones (Carter and Barham 1996; Carter et al. 1996; Thrupp et al. 1995).

Chile is a frequently cited 'success story' of non-traditional agro-export competitiveness and a good example of the productive imbalances typically at the core of this growth. As noted, one of the first orders of business for the Pinochet dictatorship was to pursue a counter land reform. The state farms that had just been established were privatized and landholding became progressively more unequal under Pinochet's rule, though not quite to the pre-land-reform levels of the 1950s. The reconcentration of land solidified a medium- and large-scale landowning class that was committed to expanding exports, and this formed a key part of the pioneering neoliberal economic model that was pursued. A remarkable growth in agro-exporting ensued: in the two decades following the 1973 coup agriculture rose from 2 per cent of Chile's total exports to almost 28 per cent. Where higher-value new exporting possibilities were created, especially with fruit farming in the bountiful Central Region, many large enterprising farmers bought land from smaller farmers who lacked capital or credit to make the necessary transitional investments (Kay 2002).

In addition to the reconcentration of land, agricultural labour became much less organized and stable than in the Allende era, and large fruit farmers were able to capitalize on labour surpluses in the small farm economy, particularly of women, to source their workforce cheaply and seasonally. As in Brazil, agro-export growth also led to intensified pressure on the environment, as deforestation, soil erosion and the toxic burden of farming all increased (ibid.). Between 1990 and 2000, the annual volume of pesticides imported to Chile more than doubled, topping 15,200 tonnes (Ross 2005). Thus, to tout Chile as a model worthy of emulation one must either 'ignore or downplay the underside and high social costs of this transformation, such as the repression of labour, the [land] counter-reform, the increase in rural poverty, the rising inequality and the environmental damage' (Kay 2002: 488).

When small farmers have been included in new agro-export models in developing countries it has tended to occur through contract farming arrangements. Advocates like to portray contract farming as a 'dynamic partnership' of freely entered and mutually beneficial contracts between small farmers, who are often short of capital, and agro-TNCs, domestic firms or large commercialized farmers whose capital, expertise and marketing capacities 'trickle down' to small farmers. In this view, small farmers can benefit from the confidence of having an assured market and price and can modernize through enhanced access to agro-inputs, credit and extension expertise. At the same time, corporate investment in agriculture brings benefits as contract farming mitigates worries about land reform, labour insurrection, new forms of regulation and welfare obligations to workers. But for critics, such an account badly obscures the limited alternatives facing many small farmers, the asymmetries of power as contracts are set (which are reflected in terms heavily favouring the buyer) and the ways in which debt, high-input farming methods and market obligations can become entrapping over time, while the overall profits in commodity chains are concentrated in distribution, processing and retail. Thus, the case against contract farming interprets it as a relationship in which capital transfers the production risks inherent in agriculture (e.g. pests, disease and drought) to farmers while restricting their share of the value within commodity chains and extending dependence on external inputs, hence increasing farmers' costs and debts and restricting their autonomy. In so doing, contract farming is seen to effectively transform small farmers into a propertied but semi-proletarianized, piecework-driven and unorganized labour force (Little and Watts 1994). Between these polar interpretations, others urge caution in generalizing about the nature of contract farming, arguing instead that the relationship takes different forms that are often full of grey areas and require careful localized exploration (Kay 2002).

Structural adjustment and agriculture

The divergent agrarian transformations in developing countries since the Second World War have been shaped by different colonial legacies, land reform struggles, successes and setbacks, state developmental strategies and the activities of national and transnational capital. Whether these led towards increased food import dependence, a push for national food self-sufficiency or competitive agro-exporting, after the early 1980s these various paths cannot be understood in most developing countries without reference to a generalized political economic context and a formulaic policy prescription, that of debt and structural adjustment. After the early 1980s, most non-oil-producing developing countries were compelled by debt into the arms and policy dictates of the IMF and the World Bank, which restructured the domestic policies of debtor countries to an extent 'which matched, perhaps even exceeded, the direct administration of bygone colonial governments' (Hoogvelt 1997: 167). This restructuring was a crucial precursor to the multilateral institutionalization of trade liberalization in the 1990s.

The context of adjustment[4] The IMF and World Bank were established at the Bretton Woods Conference in 1947 with the IMF envisaged as the global lender of last resort, tasked to prevent the sort of spiralling contractions that occurred in the 1930s Great Depression, and the World Bank to facilitate the reconstruction of Europe. In the two-decades-long global economic boom that followed the Second World War the IMF was a relatively minor international institution while Europe's quick recovery turned the World Bank's attention to the decolonizing South. Both the IMF and the World Bank have been dominated by the world's leading capitalist countries since their inception, especially the USA, and hence subject to the prevailing economic orthodoxy there. This meant that the World Bank's policies from the 1950s to the 1970s largely reflected a Keynesian mindset, with the objective of augmenting investment in major infrastructure projects like dams, irrigation schemes, industrial ports and highways by guaranteeing loans for private capital, collaborating in shared funding arrangements and directly funding loans and grants. From India to the Amazon, the World Bank had a role laying some of the infrastructural groundwork for modernizing agrarian change.

It was not until the 1970s that the World Bank's role as an infrastructure financier would begin to change, and the IMF's influence as a crisis-stage creditor and policy-maker would begin to grow. The vulnerability of narrow commodity-dependent economies across much of the developing world, which had been partially masked by strong demand and prices during the post-war boom in the 1950s and 1960s, was exposed after the onset of global recession in 1972/73, the oil price shock and declining terms of trade for tropical commodities, leaving much of the developing world with severe balance of payments problems. A common response was to turn to private US capital markets that were eager to reinvest the flood of 'petro-dollars' they were receiving (the invested earnings from oil-producing nations, much of which ended up in US banks) and saw developing-country governments as seemingly reliable borrowers. But the combination of this heavy borrowing coupled with rising interest rates and widening balance of payments deficits proved to be a toxic combination which ultimately erupted in the debt crisis of the early 1980s, with a number of major debtor nations teetering on the brink of bankruptcy.

With US financial interests so overextended, the prospect of these defaults was a mightily destabilizing one. The roles played by the IMF and the World Bank grew and began to blur as they helped stave off this crisis in the global financial system by plugging the massive payments problems in the developing world with loans that were accompanied by the interwoven series of conditionalities contained in SAPs. After the debt crisis, the IMF used protracted payment imbalances to legitimize its recurrent lending in many countries, which entailed a far deeper and more durable role than its original crisis mandate, while the World Bank began providing broader loans that were no longer tied to specific projects and were instead connected with IMF conditionalities. These conditionalities were further engrained through bilateral (government-to-government) lending and aid, particularly from the United States Agency for International Development (USAID), and in the Americas through loans from the Inter-American Development Bank.

Around this time the political rise of neoliberalism, marked by the victories of Ronald Reagan in the USA and Margaret Thatcher in the UK, was having a major impact on IMF and World Bank policies, with the US Treasury dominating the IMF. SAPs disguised deeply ideological assumptions under the cloak of a benign inevitability – a theoretical faith likened to a religion with the notion of 'market fundamentalism' – best captured by Thatcher's assertion that 'there is no alternative'. As a result, there was little contextual sensitivity and a similar package of reforms was stamped upon every debtor nation, generally including: trade and investment liberalization; export promotion; currency devaluation; fiscal austerity; price and wage deregulation; the privatization of state services and enterprises; and the assurance of private property rights.

Given the location of its architects, this prescription came to be known as the 'Washington Consensus' (Williamson 1990), which hints at the implicit political agenda underlying the explicit economic rationale. Bello (2002a, 1994) argues that the fervent, crusading Reagan administration loathed the rising assertiveness of developing countries and the calls for structural change in the global economy occurring in the 1970s (e.g. G-77, NAM, NIEO, UNCTAD) and found in the debt crisis essential leverage through which to constrict the sovereignty of restive developing countries and put them back in their place as externally managed suppliers of cheap commodities. Thus, according to Bello, adjustment represents a backlash of sorts, designed to debilitate meaningful political collaboration between developing countries and their ability to construct nationally and regionally integrated economies or producer alliances capable of supply management such as the Organization of Petroleum Exporting Countries (OPEC). In addition to the leverage of debt, the spread of SAPs was also facilitated by the fact that the policies meshed with the interests of national elites in developing countries and that many government officials and economists were trained in the USA and Europe and immersed in its economic orthodoxy.

Structural adjustment in theory The broad theoretical basis for adjustment rested on an interpretation that ascribed trade imbalances and debt in developing countries to the 'irrational' or inefficient allocation of capital, resources and labour, caused primarily by inflated and often corrupt public sectors and state interference with market forces (e.g. excessive tariffs, regulation and taxation, public ownership and/or management of industries and services, etc.). So by rolling back the state and increasing the discipline of the global market, which is framed as the ultimate arbiter of economic 'rationality' because it does not tolerate inefficiency, the belief was that adjustment policies would force capital, resources and labour to shift within or between economic sectors in line with the reified law of comparative advantage (i.e. towards the most internationally competitive industries or firms). Further, this would facilitate access to cheaper and more diverse imports and provide 'transfer benefits' to developing countries in the form of increased capital, skills and technology. Together, the input efficiency of the economy as a whole was expected to improve, enhancing export competitiveness and the ability to generate foreign exchange and service foreign debts, which would in turn ultimately correct macroeconomic imbalances. Leys (1996: 24) sums this approach up concisely: 'it is hardly too much to say that by the end of the 1980s the only development policy that was officially approved was not to have one – to leave it to the market to allocate resources, not the state', with social costs washed away in abstraction and the promises of the market.

One of the most elemental assumptions of SAP policies for agriculture was that freer markets enhance food security. Because global market integration exacts pressures for production everywhere to be defined by the rationality of comparative advantage it is expected that integration will lead the global food economy as a system towards increasing efficiency, maximized production and stabilized low prices and supply. This last point is summarized by Winters' (1988: 20) insistence that 'self-sufficiency offers no direct benefits to a country in and of itself ... the world market will almost always be more stable than any single country could be'. Following this logic, for a nation to optimize its own food supply it is expected to shift resources into internationally competitive sectors, agricultural or otherwise, increase its export revenues and buy on the world market those items produced more cheaply elsewhere, as in the claim that 'for food security, a nation's agriculture matters more as a provider of individual buying power and earner of foreign exchange' (Tweeten 1997: 245). With food security a function of international competitiveness and the social and cultural significance of agriculture fading into the background, liberalization is then normalized and it becomes 'irrational' for a nation to support or protect uncompetitive domestic production.

SAP policies for agriculture were also heavily influenced by the literature critiquing the urban bias of development planning, which argued that agriculture was subsidizing urbanization and industrialization (or, conversely framed, that cities were acting as 'vampires' on their countrysides) and mistakenly drawing resources and investment out of farming (Lipton 1977). Adjustment policies attempted to address this by reducing state interference in markets (e.g. price controls), cutting food subsidies and transferring state marketing boards to ostensibly more efficient private actors in the belief that these changes would reduce marketing costs, improve flexibility and increase farm-gate prices, thereby stimulating production (Badiane and Kherallah 1999). Another policy seen to have propped up industry at the expense of agriculture was the overvaluation of exchange rates, which indirectly subsidized food imports and protected inefficient manufacturers while acting as a further tax on agro-exports in addition to the often already sizeable direct taxes on exports. Thus, currency devaluation (dubbed 'getting the prices right') and the reduction of export taxes were expected to make agro-exports more competitive. Devaluation was also anticipated to make production more competitive in domestic markets, in principle creating price incentives for farmers to increase production, either ignoring or downplaying the role that trade liberalization and the lowering of agricultural tariffs and quotas would have. While agricultural development programmes were limited both directly by conditionality demands for reduced state spending and indirectly by debt-induced fiscal shortages, government supports that remained were to be focused on expanding traditional and non-traditional agro-exports. Export promotion was, however, to be led by foreign and domestic capital motivated by favourable incentives such as low taxes, wage controls and minimal environmental regulations. USAID also had an important role in development projects aimed at enhancing non-traditional exports, particularly within the Americas (Thrupp et al. 1995).

Finally, while the IMF and the World Bank used their great leverage to enact many difficult policies they conspicuously avoided redistributive land reform, the 'most important of all desirable "structural adjustments"' (Berry 1997: 11). In delegitimizing the state's redistributive function and sanctifying existing property rights, SAPs effectively set the highly uneven landscapes of the developing world beyond the realm of state intervention and agrarian policy. The lack of institutional support for redistributive land reform was especially striking as World Bank researchers and publications came to repeatedly advance the theoretical economic merits of a smallholder landscape rooted in the case that there is an inverse relationship between farm size and productive efficiency (Deninger and Binswager 2002; Binswager and Deininger 1997; Cornia 1985). The only action undertaken in the name of land redistribution was to be market-based and 'non-confrontational' (Leonard and Manahan 2004), which generally fell under the banner of land market reform – essentially a set of policies affecting land taxation, the costs of land market transactions (absorbing them in land banks) and/or the accessibility of long-term capital (through mortgage banks). But as Carter and Coles (1998: 169) put it in economic terms, 'the expectation that land market reform policies can shift land to the rural poor, by facilitating interclass land market transactions, presumes that the rural poor do not suffer a fundamental competitive disadvantage in the sphere of production and marketing that affects their potential to participate in the land market'. In non-economic terms, the suggestion that a land market could be a neutral intermediary and make property rights more equitable ignores how markets work in capitalist societies.

Structural adjustment in practice Many large developing-country debtors could indeed have been accused of corruption and misspending, from the rotating military governments in Brazil, Argentina and Nigeria, to the PRI in Mexico, to dictators like Suharto, Marcos and Mobutu in Indonesia, the Philippines and Zaire. Further, in pursuing import-substitution industrialization many of these governments overpaid for foreign assets and continued in dependent relations with TNCs (e.g. management contracts, technological rents), while failures to address large social disparities limited domestic markets for the manufactured goods. But adjustment planners failed to recognize that commodity dependence, unequal exchange and declining terms of trade were important roots of the debt crisis. As a result, the prescription of liberalizing markets, competing according to comparative advantage (which typically meant deepening narrow colonial ruts), and rolling back the state not only did not tend to resolve persistent trade imbalances and payments problems but in most instances exacerbated them, and as interest rates rose so too did the cost of loans.

Chang (2003) likens SAP conditionalities to 'kicking away the ladder' for constricting the ability of developing countries to set the terms of their engagement with the global economy, such as protecting domestic producers and infant industries, regulating capital markets and designing investment policies to maximize the transfer benefits derived from TNCs. That is, they denied the use of policy mechanisms that had been central to East Asia's state-supported industrialization, as well as to Western countries such as the USA at earlier stages of their development. In this view, forcing liberalization upon weak, unindustrialized economies is either naïvely ahistorical or utterly duplicitous. Compounding this, the deregulation of capital markets often precipitated the significant flight of elite capital as greater and less risky returns could be found in financial institutions in the USA and Europe than in productive investments in their own countries. The net result of adjustment prescriptions amid the unevenness of the global economy was that loans typically begat more loans and more conditionalities, staving off defaults as economies moved from one crisis to the next, while debts expanded and state resources withered. This exacted enormous social costs as funding for public goods like health, education and public housing shrank and some basic services were privatized (with the denationalization of state assets, often at cut-rate prices, providing lucrative investment opportunities). Because of this, critics have argued that SAPs provided insurance on the returns of overextended US financial interests while socializing the costs on to to the poor of the developing world through austerity demands, in effect transferring the burden of the global financial crisis.

To appreciate the impact of adjustment on agriculture the starting point is the issue of land. Agricultural adjustment began with the assumption that existing property rights are a given, however inhumanely they may have been constructed. As noted, the only land distribution policy the World Bank has supported in some instances (e.g. in Brazil, Colombia, Cambodia, the Philippines, South Africa, Thailand) is land market reform, and this has failed to dint the dualism of land ownership as extensive landowners are not often willing to sell and small farmers and large landlords are typically in exceptionally unequal financial and bargaining positions (Leonard and Manahan 2004; Borras 2003). As a result, rather than a means to redistribution these reforms should be understood as being principally aimed at diffusing 'the popular demand for radical land reform, while seeking to maintain the confidence of existing landowners and potential investors in the existing framework of land ownership' (Leonard and Manahan 2004: 9). The regressivity of the adjustment approach to land was compounded in some instances by the fact that policies designed to promote agro-export expansion favoured large commercialized farmers. Further, when small farmers took on increasing debt loads in attempts to participate in new, higher-value agro-export networks, for instance through contract farming, these became 'a major cause of land loss among the marginalized and poor sectors of society', with larger landowners able to capitalize on distress sales by 'buying up land at deflated prices' (ibid.: 5).

In addition to the absence of redistributive land reform, a number of other constraints inhibited small farmers from benefiting from the promises of adjustment's new incentive structure. In terms of non-traditional agro-export growth, only a few developing countries succeeded in widely moving into higher-value crops such as fresh fruit and vegetables, which might conceivably bring greater per-area returns to small farmers, and, as discussed, there have been serious barriers to small-farmer participation in these markets. These barriers were exacerbated by the fact that the state capacity to support small-farm programmes through such things as extension services, input subsidies, low-interest credit and public research initiatives was eroded by debt and adjustment. The developmental case for increased traditional agro-export promotion was even more dubious. With adjustment occurring in the context of steadily falling prices for the traditional agro-exports, more production was needed simply to maintain the same earnings, inducing a vicious cycle of overproduction, structural oversupply and downwardly spiralling prices. Coffee was an especially disastrous case, as adjustment encouraged expanded production in various countries as major new producers entered the global market (Vietnam and Ethiopia) and the multilateral attempt at supply management broke down; locked into a state of oversupply, by 2002 the inflation-adjusted world market prices of coffee had fallen to 14 per cent of what they were in 1980. While coffee was an extreme case it was hardly unique, with 2002 adjusted world market prices but a fraction of 1980 levels for such tropical commodities as cocoa

(19 per cent), cotton (21 per cent), palm oil (24 per cent), groundnuts (38 per cent) and tea (47 per cent) (Robbins 2003). This problem was exacerbated by continuing selective protectionism against processed commodities in rich countries and agro-food TNC control over distribution and marketing.

While 'getting the prices right' by deregulating markets and devaluing currency was expected to improve farm-gate prices for small farmers and enhance their competitiveness vis-à-vis imports, the costs of production also rose because of declining public investment in infrastructure and small-farm programmes and the fact that production costs also rose as devaluation made imported inputs more expensive. Further, the hypothesized gains to small farmers from privatizing state marketing boards often failed to materialize, especially where middlemen could parlay their leverage into higher margins for themselves in the absence of regulation or much competition, particularly problematic in remote areas. So even though market prices were not to be held down in domestic markets and devaluation made imports relatively more expensive, many small farmers were not more competitive in this new context as imports rose. As a result, 'getting the prices right' repeatedly led to increasing food costs but not increasing earnings for farmers, with the social fallout from this borne especially by the urban poor, who were dependent on cheap food. The shock of the initial price rises was a key spark in what became known as food or IMF riots (Walton and Seddon 1994) or what the World Bank (2001) called 'austerity protests'.

The imbalances associated with SAPs and agriculture also relate to how the abstract logic of the free market approach to food security played out in reality. At the same time as adjustment was forcing developing countries to be more open and export oriented and cutting state support for agriculture, the USA and Europe were intensifying their subsidy-fuelled export competition and selectively increasing protectionism with tariff and non-tariff barriers – devastating the logic's core assumptions about the neutrality and efficiency of comparative advantage. The average support given by governments in OECD countries to their agricultural sectors increased from 32 per cent of the total value of agricultural output in 1979–81 to 47 per cent in 1986, and nearly every OECD country increased its agricultural protectionism against processed or directly competitive food over this period (Burniaux et al. 1988). These large subsidy levels, which were entwined with the overproduction, deflated prices and export imperatives in the temperate grain-livestock complex – and which continued increasing into the mid-1990s, especially in the EU and the USA – lay waste to the benign image of comparative advantage that was given to the global marketplace in the rosy theory of adjustment planners. In addition to ignoring the subsidy-fuelled market distortions at the heart of the global food economy, the SAP image of comparative advantage ignored the arguably even bigger environmental subsidies (i.e. externalized costs) underwriting this system, as discussed in Chapter 1. In short, where SAPs fostered increased food imports, new consumption patterns and reduced self-sufficiency in developing countries this could hardly be ascribed to the simple playing out of comparative advantage on a level field.

Finally, since small farmers tended to be largely or wholly excluded from new agro-export networks, the selective protectionism in the industrialized world had little effect for them. But it was very frustrating to the increasingly competitive commercial class of farmers that had emerged, especially in the new agro-export powerhouses. This frustration was reflected in the criticism of the then president of the Inter-American Development Bank 'that the farmers of Latin America could compete against their counterparts in the industrialized countries, but not against the treasuries of these countries' (in Rampersad et al. 1997: 131). As will be discussed in the following chapter, this argument has been a recurring one in the struggle over the multilateral regulation of agricultural trade.

Conclusions

Amid all the revolutionary changes in global food production, most of the farming population in the developing world still continues 'hoeing and ploughing in the old, labour-intensive manner' (Hobsbawm 1994: 293), typically confined to small plots within highly uneven landscapes. In spite of the tropical commodities disaster, agro-export production continues to command a disproportionate share of the best arable lands across much of the developing world, while the establishment of competitive, industrialized agro-export platforms opened new opportunities for accumulation and set further barriers to long-repressed movements for land reform in some instances. Increasingly these small farmers are languishing under deflated prices for basic food crops as markets integrate, greatly influenced by large-scale, industrialized production from the USA, Europe, Canada, Australia and New Zealand as well as from the new agro-export powerhouses in the developing world. Debt and adjustment have compounded the land inequality and price squeeze throughout most of the developing world, as the state virtually ceased to exist in any meaningful, supportive way for small farmers (e.g. for land distribution, extension, research, credit), if it ever did previously. The many sub-Saharan African nations that remain bound to the traditional export/cheap grain import model have fared the worst, as recurring famines and persistently high levels of undernourishment attest.

Thus, many of the world's poorest farmers are facing a new vulnerability, having been incorporated into markets in which they struggle to earn incomes that can sustain their households. As discussed in Chapter 1, this is part of a contemporary social convulsion that poses especially difficult food security questions as urban poverty mounts, and could well end up reinforcing cheap food policies if concentrated, frustrated, hungry urban populations are conceived as a more politically influential constituency than are seemingly 'redundant' small farmers. But the counterpoint to this is the fact that peasant farmers have historically been at the forefront of revolutionary change in many parts of the developing world in the past (Chaliand 1977; Huizer 1973; Wolf 1969) and their submission or exhaustion in the face of gravely inequitable national and global inequalities certainly cannot be assumed (Veltmeyer 2005; Desmarais 2002; Bernstein 2000).

With such large and pressing questions bound up in the trajectory of the global food economy, agency in affecting its course has shifted steadily away from states, communities, farmers and democratic institutions, and more and more towards agro-TNCs. Structural adjustment went hand in glove with rising agro-TNC consolidation throughout the global food economy, as markets were opened, their ability to source production was enhanced and dependence on corporate agro-input packages grew. Yet though well served by the bilateral reforms imposed by the IMF and the World Bank, in the 1980s agro-TNCs began seeking a supranational authority that would further secure and expand market access and facilitate sourcing flexibility. As McMichael (2000a: 139) puts it: 'behind the apparent multilateralism of the World Trade Organization (WTO) stands the attempt to institutionalize rules of a neoliberal world order to match (and deepen) the corporate-led economic integration underway'. It is here that attention now turns.

Notes

1 The assumptions, practices, discourse and impacts of development theories have been thoroughly analysed in a massive literature that lies beyond the realm of discussion here. See Leys (1996) for an especially good critical review.

2 From the 2006 home page, accessed at: <www.yum.com/about/>. © 2006 Yum! Brands, Inc.

3 This event was described to me by a friend who attended the rally.

4 This summary discussion on the context of adjustment and the nature of structural adjustment draws especially from SAPRIN (2004), Bello (2002a, 1994), Stiglitz (2003), Petras and Veltmeyer (2001), Chossudovsky (1998) and Hoogvelt (1997).

FOUR

Entrenching an uneven playing field: the multilateral regulation of agriculture

The rise and imbalances of the WTO

The WTO's Agreement on Agriculture (AoA) came into effect in 1995 and constitutes a major landmark in the development of the global food economy as it set in place, for the first time, multilateral rules restricting the sovereignty of governments to establish their own agricultural policies. The officially stated objective was to establish 'a fair and market-oriented agricultural trading system', and the heading of the WTO website for agriculture boldly proclaims: 'Fairer markets for farmers.'[1] It would be hard to dispute that some disciplines on agricultural trade in the interests of 'fairness' were not needed at the time the WTO was being constructed, given such things as the distortions caused by rich-country surpluses and subsidies, the declining earnings of farmers within commodity chains, the instabilities associated with rising import dependencies, the tropical commodities disaster and the bilateral policy restructuring of structural adjustment occurring in most developing countries. The form of regulation undertaken in the WTO's AoA, however, bears no relation to an exercise in promoting fairness – 'fair' instead acts as a soothing, ideological pretext for 'market-oriented', and 'market-oriented' as a neutral-sounding way of framing the fact that the legal rights of agro-TNCs were advancing vis-à-vis the capacity of states and local governments. The aim of this chapter, then, is to explain the significance of the AoA with respect to the trajectory of the global food economy described hitherto in the book, focusing on the essential disciplines, imbalances and politics of this rules-making without tracing all the WTO's Byzantine legal details (though in the instances where 'the devil is in the details' some technical discussion is necessary), which will ultimately seek to illuminate why the struggle to extend, transform or terminate this agreement emerged as such a critical dimension in the battle for the future of farming.[2]

Because the struggle over the AoA cannot be entirely separated from other disputes about the WTO, as it is just one especially heated conflict among a number of others (each of which has both individual and collective dimensions), it is important to begin by clarifying how agriculture fits within the evolution and framework of the WTO. The roots of the WTO lie in John Maynard Keynes's vision, in the wake of the Great Depression and the Second World War, for a system of global economic governance to manage the cyclical nature of capitalist economies. This system was to have three institutional pillars: the IMF, the World Bank and a body to promote trade and defend against protectionism. But the embodiment of this third pillar – the General Agreement on Tariffs and Trade (GATT) – was much weaker than Keynes had envisioned as many nations failed to sign on and its scope was limited to reducing tariffs on trade in merchandise goods while other sectors such as agriculture remained outside of GATT disciplines. In the case of agriculture, the same economic, social, cultural and strategic imperatives that had fuelled the enormous subsidy regimes in rich countries kept it from multilateral trade discipline. It was not until the Uruguay Round negotiations (1986–94) of the GATT that a broader legal framework for global trade was again seriously considered, and the WTO was ultimately born out of these long negotiations, not only drawing in other commodity sectors like agriculture but widening the boundaries of trade and investment regulation to include such things as the General Agreement on Trade in Services (GATS), Trade-related Investment Measures (TRIMS) and Trade-related aspects of Intellectual Property (TRIPS).

The WTO is essentially a set of shared rules about the degree to which governments can protect and subsidize domestic economic activity, a judiciary to enforce commitments and a forum to recurrently draw countries together to rework these rules. The rules are negotiated by sector and hence contain different specific commitments, but they are guided by common principles that essentially fall under one overarching objective: to entrench and extend the rights of transnational capital in trade and investment by reducing so-called 'unnecessary barriers to trade' and 'discriminatory' trade practices of governments. Meanwhile, the long-standing trade principle of 'special and differential treatment' (SDT) – which assumes that the colonial malformation (i.e. the dependence on a narrow range of commodities) and often small size of developing-country economies give them particular rights in trade agreements – was significantly diluted in the WTO. SDT had previously been defined in perpetuity (described as 'hard' SDT) in agreements such as those between the EU and ACP (Africa, the Caribbean and Pacific), but under the WTO it was reduced to giving developing countries longer time frames to adjust to commitments ('soft' SDT). This implied a huge shift in how trade is understood, away from something that needs to be managed to serve a developmental role and towards a view where liberalizing trade is in itself portrayed as a development policy.

Once negotiated sector by sector, the end product is tabled as an all-or-nothing prospect for governments, meaning that they cannot pick and choose agreements; countries can still also be part of other bilateral and regional trade agreements but these must be 'WTO compatible'. The commitments made in the various sectoral agreements are then backed up by a dispute settlement process in which aggrieved governments – when they feel their national economic interests have been injured by another country's non-compliance, and commonly at the behest of corporations or whole industries – can take their case to the WTO's Geneva-based Appellate, where it is then adjudicated by a panel of trade lawyers. The panel decisions can be appealed but not vetoed by governments, and if the violating country does not come into compliance the judiciary has the authority to inflict economic penalties in the form of fines and trade sanctions. While the rules enforcement process is non-elected and non-transparent, one of the great illusory promises of the WTO was that the legislative process was a genuine multilateral democracy (unlike the IMF and the World Bank, which are openly governed by unequal voting ratios that ensure US and European control) in which consensus was to be sought in ongoing negotiations and each country would have one vote at the Ministerial General Council and was free to sign on or not. Then, once these rules were made, each country would have the same recourse to the technical, non-partisan WTO Appellate.

But in practice there is a deep divide between the formal process and informal power structure of the WTO, as the Uruguay Round and the Ministerials that followed made a mockery of the rosy image of a forum of equal partners. As Bello (2002b: 119) puts it, '"Consensus", WTO style, means the big trading countries impose their consensus on the less powerful countries.' The Uruguay Round was basically forged by the USA, the EU and to a lesser extent Japan, with power imbalances playing out through such things as the outright exclusion of many countries from key meetings, the back-room influence of corporate lobbyists and the extreme asymmetries in the size of developing-country negotiating contingents that were magnified by the voluminous legal documentation. Bello also emphasizes how it would be naïve to suggest that developing countries were merely duped, as elites at the helm of their export economies have exerted a great deal of political influence and typically view greater market integration with sincere ideological commitment, self-interest or both.

Further, even those governments with more cautious attitudes towards market integration faced the systemic compulsion of global capitalism in which the prospect of being *excluded from* transnational flows of trade and investment might appear as being a greater risk than being *exploited within* an uneven system, with fears of exclusion making abstention from the WTO seem a daunting prospect. Yet despite the export-biased interests that developing countries brought to the negotiating table, the North–South power imbalance was such that little attention was given to the trade concerns of developing countries, such as the tropical commodity disaster and the selective protectionism of rich countries against processed and semi-processed goods, to say nothing of the interests of poor farmers in the face of rising food imports and deflated prices.

The push to regulate global agricultural trade under the WTO

While national agricultural policies had been kept from multilateral trade rules prior to the Uruguay Round, by the 1980s a few major pressures had emerged to bring agriculture to the negotiating agenda. First, the rise of neoliberalism in the USA and Europe led some governments to question the fiscal burden of their subsidy-fuelled agro-export competition, as agriculture had become by far their most heavily state-supported economic sector (provided that neither the US military-industrial complex nor the unaccounted environmental costs are understood as subsidizing the oil industry). Because both sides were also unwilling to address the structural overproduction fuelling their export imperatives, large, unilateral subsidy reductions were not seen to be possible without seriously impairing the competitiveness of their farm sectors abroad. In this regard, the USA was particularly interested in scaling down European export subsidies.

Second, a group of middle- and smaller-scale agro-exporting countries, with competitive but less subsidized agro-export sectors, formed an advocacy association to amplify their calls for increased trade liberalization as well as to lobby for the scaling back of the USA and the EU subsidy regimes, as these subsidies were seen to impede their export growth both to third-party markets and to the USA and the EU themselves. Known as the 'Cairns Group' for the site of their first meeting, this association drew together developed countries Australia, Canada and New Zealand with a number of developing countries with competitive agro-export platforms whose political advocacy was squarely on the side of their commercialized farming classes, which were interested in market access issues (Argentina, Brazil, Chile, Thailand, South Africa, Uruguay, Bolivia, Colombia, Costa Rica, Fiji,

Guatemala, Indonesia, Malaysia, Paraguay and the Philippines). The Cairns Group served to add a significant new voice to calls for market liberalization but their demands on subsidies were largely deflected, which came to be a simmering source of tension.

The third and most decisive factor pushing agriculture into the Uruguay Round was the pressure exerted by agro-TNCs through US and European governments. Governments acting in the interest of business elites is of course nothing new – Marx had long ago highlighted this as a basic dynamic of capitalist societies – and while it might not have been framed as bluntly as 'what is good for Monsanto is good for the rest of America',[3] variants of this old-fashioned reasoning, however spurious, still helped to mask how things like political contributions, lobbying and revolving doors between public and private executives greased this state–capital nexus. As discussed in Chapters 2 and 3, agro-TNCs were pleased but not satisfied with how they had given shape to subsidy regimes and with how favourable trade and investment reforms had been bilaterally propelled upon most of the developing world through adjustment, and they began pressing for a supranational institution that would both legally entrench their control and contain the authority to extend it further. In essence, they wanted to maximize their flexibility selling and sourcing within and between nations by shifting the locus of sovereignty within the global food economy, moving significant elements of regulation beyond the legislative reach of governments (McMichael 2000a, b). The motive for expanded market access was centred on Asia, which an OECD (1999: 9) report described as 'the main driving force behind the growth in agricultural markets and trade'. With rising affluence and the rapid meatification of diets across much of industrializing Asia providing growing demand for products from the temperate grain and livestock complexes, and economically booming nations like China and especially India still relatively small importers of food relative to their enormous populations, agro-TNCs were acutely aware of the tremendous potential for market growth that institutionalized trade liberalization could unleash.

Another important reason why agro-input TNCs were eager for a supranational institution to regulate key aspects of the global food economy relates to the web of inputs in industrial agriculture and especially its linchpin, the seed, which is entwined with the rise of GMOs and the desire to project US-styled intellectual property laws. Not only does the ability to extract technological rents depend upon the legal sanctification of intellectual property rights at the point of production, and hence retaliatory recourse against farmers for saving 'enhanced' seeds and governments for failing to enforce patent protections (at least prior to the day terminator technologies proliferate, if they ever do), it ultimately also rests upon ensuring that GM products will have access to markets. If GM crops can become widespread in some agricultural systems (e.g. as they have in the USA, Argentina and Canada) but other governments (e.g. the EU, Japan) can have them prohibited or even merely insist they be labelled, it can pose a grave threat to export markets given growing consumer consciousness about environmental and health risks. Thus, in addition to attempting to shape the international food safety standards of *Codex Alimentarius*,[5] agro-TNCs have viewed multilateral trade rules as a way to override popular resistance to GMOs by legally containing trade barriers and labelling requirements – in other words, moving debates away from consumers and elected governments and locating decision-making in the shadowy world of trade negotiations and corporate lobbyists (as will be seen, this has been a very divisive issue between the dominant architects of the WTO and is a vital fault line for contestation). This ability to override popular resistance could also potentially bear on the ability to enact regulations for the more humane treatment of farm animals, something that has not been a conflictive multilateral issue on ethical grounds, but which has drawn attention from how it impinges on human health through the use of growth hormones and other pharmaceuticals. In short, agro-TNCs saw multiple rewards in a binding set of multilateral rules for the global food economy, and these were so significant that they were willing to risk some disciplines being brought to the USA and the EU subsidy regimes that had long helped to provide them with price-deflating overproduction, especially since they were confident that cheap surpluses could be obtained by pitting the world's farmers against one another and that subsidy regimes were not going to just vanish.

The role of agro-TNCs in shaping the AoA was seen most plainly in the oft-noted fact that a former Cargill executive spearheaded the US agricultural negotiating team during the Uruguay Round, and the agreement itself amounted to a compromise between US and EU negotiators who were each bent on securing the interests of their corporations and large-scale farmers. Nevertheless, developing-country governments constituted 80 per cent of the WTO's signatories, while representing an even higher percentage of the world's farmers.

The new 'discipline' for the global food economy

The commitments of the AoA are generally framed in terms of disciplines given to domestic supports, export subsidies and market access. The most important of these was enhanced market access for exporting parties. For governments, this meant losing their ability to limit imports by volume (e.g. quotas) and reduced flexibility to affect the competitive position of imports by value (e.g. tariffs), with tariffs established as the only policy tool to restrain imports, their assessment standardized and their levels then capped and to be eventually cut. The only exception to this was permission to use scientifically determined sanitary and phytosanitary (SPS) measures to protect against imports posing environmental and human health risks, with the definition of risk obviously a matter of great import. All non-tariff barriers such as quotas were to be converted into tariffs (a process known as 'tariffication'), using a formula to quantify the impact of non-tariff barriers on market prices, and no new non-tariff barriers were allowed. The standardization of tariff assessment under the WTO ensured that they were calculated as a percentage of the *invoice price at the border* for an item, when many nations had previously calculated tariffs as a percentage of the estimated *market price* under the Brussels Definition of Value; this increased market access in a subtle way because the former is typically much lower than the latter. Ambitious schedules were then set for tariff cutting which involved both a larger average, across-the-board reduction (36 per cent in six years for developed countries;[4] 24 per cent in ten years for developing countries) and a smaller minimum cut per tariff line (15 per cent for developed countries; 10 per cent for developing countries), which gave a measure of flexibility from item to item. The soft SDT given to the world's LDCs and some developing countries was that they did not have to commit to any schedule of tariff reduction at this time, but they were forced to 'bind' the upper limits for tariffs on all items into the future (also known as setting 'tariff ceilings'). Given flexibility here, many developing countries set these ceilings relatively high, above the actual pace at which liberalization was occurring under structural adjustment.

While committing to a tariff ceiling did not accelerate liberalization it did limit future flexibility and establish a highly vulnerable target for future negotiations, a point that requires some elaboration. From their position at the helm of the WTO negotiations, the USA and the EU liked to characterize it as a horse-trading system; that is, one where mutual concessions are wrestled over and ultimately exchanged. In agricultural trade negotiations, the primary policy mechanisms at play relate to border protections and subsidies. At the Uruguay Round, developed countries made larger relative concessions on subsidy discipline and tariff reduction. Most developing countries on the other hand entered the Uruguay Round with minimal agricultural subsidies, as any state support programmes they may have had for agriculture were in the midst of being rolled back by debt and adjustment conditionalities, giving them little to concede on this front; they could set high tariff ceilings and were allowed to make relatively smaller cuts to average tariffs, though these had also widely been affected by adjustment conditionalities. Thus, the USA and the EU argued that they were the ones making the main concessions, pointing especially to their subsidies disciplines, while accusing developing countries of being 'free-riders' for getting the competitive benefits of this new discipline while making relatively lighter concessions. So while slower tariff reduction schedules and high tariff ceilings were permitted in the original AoA it was clear these would be targeted at future WTO Ministerials, with developing countries holding few other bargaining chips, especially after the bilateral policy changes wrought by structural adjustment.

This leads into the other two elements of the AoA's new regulatory discipline, export subsidies and domestic supports – in other words, to the very uneven realm of agricultural subsidy regimes. A key part of the argument that the AoA was creating a more 'fair' regulatory framework for agriculture was that increased market access was being accompanied by efforts to reduce the competitive distortions caused by the subsidies in the world's agro-exporting giants, the USA and the EU. Because governments were losing the flexibility to protect national production from external competitive pressures, a corollary was that the impact of subsidies on the cheap surpluses deflating world market prices had to be mitigated. It is significant to note that the theoretical case for disciplining subsidies was made not only by more progressive critics of the global food economy (who were not against government supports for agriculture per se, just their current nature) but at the other end of the ideological spectrum by advocates of neoliberal trade theory, who generally see government intervention as disruptive to the competitive functioning of markets and inconsistent with the rationale of liberalization and comparative advantage (Johnson 1998). The significance of this becomes evident later on in the debate between opponents of the WTO about what should be the ultimate objective of this opposition.

With the theorized case for the liberalization of agricultural trade so intimately connected to subsidy discipline, the nature of the discipline given to US and EU subsidies was a crucial part of the AoA's moral appeals to fairness. The bulk of agro-subsidies were deemed permissible under the AoA, however, meaning that they could not legally be used as grounds to protect against agro-exports from the subsidizing country, something that would become central to subsequent disputes over the AoA. Given that the fiscal burden of agro-subsidies was one reason why agriculture had been brought to multilateral trade negotiations, this might seem a strange outcome. The political interest in cutting subsidies did not occur in a vacuum, however, and with both the USA and the EU perceiving the economic and strategic importance of agro-exports and beholden not only to corporate but also to big-farmer lobbies, their negotiators ended up crafting a nuanced defence of subsidies. This defence, both at the Uruguay Round and beyond, was accompanied by a rather cynical exercise of cross-Atlantic finger-pointing, in which each side deflected criticism about their own subsidy regimes (vocalized especially by the Cairns Group) by rebuking aspects of the other, and in the process ratcheting down commitments.

In technical terms, the defence of agro-subsidy regimes hinged on a limited definition given to what were defined as problematic or 'trade-distorting' subsidies that would then be subjected to reduction commitments, with rich countries committed to relatively faster and greater reductions than developing countries. Export subsidies were naturally seen to be the most trade-distorting, with some forms prohibited and others put in the 'amber box' (a traffic light analogy for 'slow down') and given the toughest reduction commitments. The cuts to export subsidies in the amber box were based on average annual levels between 1986 and 1990 and specified according to both their total value (36 per cent over six years in developed countries and 24 per cent in developing countries over ten years) and to the volume of subsidized exports (21 per cent over six years in developed countries and 14 per cent over ten years in developing countries), again with no commitments for the world's LDCs, which didn't have these to begin with.

Beyond export subsidies, the matter of what was seen to distort trade and then subjected to reduction commitments gets more complex, because the bulk of the subsidy regimes are not export subsidies but rather domestic supports. Under the AoA, domestic supports were classed in either amber, green (for 'go ahead') or blue boxes. The domestic supports put in the amber box were those considered to send artificial signals – as opposed to the 'natural' signals sent by markets – that encourage overproduction and hence distort trade, both through surpluses that end up dumped in global markets and as a barrier to import competition. These involve payments directly 'coupled to' production; that is, where increased production brings increased subsidies to farmers, such as price supports (a type of subsidy that corporations have long opposed). The discipline for domestic supports in the amber box started by aggregating both their crop-specific and non-crop-specific forms through what is called the 'Total Aggregate Measurement of Support' (total AMS). An annual average of total AMS was taken from the base years 1986–88, with developed countries committed to cutting these levels by 20 per cent over six years and developing countries committed to 13 per cent cuts over ten years. A maximum baseline for amber box domestic supports was allowed, measured relative to the total value of agricultural production and set at 5 per cent for developed countries and 10 per cent for developing countries, and it was promised that future Ministerials would address further reductions to the domestic supports in the amber box. Again the LDCs, which did not have significant supports in the first place, were exempted from commitments in the Uruguay Round as part of the soft SDT, though they were prevented from raising these forms of support above specified levels into the future.

While the claim was that the different boxes distinguished between the supports that have bad outcomes on production (stimulating it beyond market signals) and trade (distorting comparative advantage) and hence require discipline, and those that are largely production neutral and non- or minimally trade-distorting investments and hence do not, in reality the impact of subsidies is much more of a grey area. And most of this complexity gets washed away in the green box, which was exempted from reduction commitments and countervailing duties, while these definitions and subsidy levels were cast beyond legal challenges until 2003 by the curiously named 'Peace Clause'. The green box embraces a large range of subsidies, including state support for: research; pest and disease control; infrastructure; food aid programmes for domestic consumption; inspection services; marketing promotion; public stocking for food security; disaster relief; crop and income insurance schemes; and some forms of direct payments to farmers for income support, restructuring production, environmental initiatives and regional assistance that are 'decoupled from' current production or prices. Subsidies classed in the last category, the blue box, were also exempted from disciplines in the AoA. The blue box included supply-limiting programmes such as payments made to take farmland out of use for conservation objectives, which were long targeted by agro-TNCs to the point of complete elimination in the USA, and those that support agricultural and rural development in developing countries, which were generally small to begin with.

Finally, though it is separate from the disciplines contained in the

AoA, another key way in which agro-input TNCs sought to insulate their control within the global food economy through the WTO was with the TRIPS protocol. This was also spearheaded by US negotiators to force governments to enact stronger patent protection laws, premised on the claim that the prospect of earning an exclusive technological rent for a specified time period must be protected in order to guarantee incentives to invest in research. A Monsanto executive encapsulated the pressure behind TRIPS in a frequently cited quotation: 'Industry had identified a major problem in international trade. It crafted a solution, reduced it to a concrete proposal and sold it to our own and other governments.' By advancing the legal protections for scientific innovation as commodities, which concurrently erodes the conception of innovation as a public good, and advancing the reach of what could be considered intellectual property to include such things as manipulated genes, TRIPS helped to secure the ability of agro-input TNCs to compel farmers into the market for their seeds, recalling the crucial role of the commodified seed in the corporate-controlled web of inputs in industrial agriculture.

Unpacking the imbalances: from contestation to a crisis of legitimacy

Not long after the WTO was unveiled – literally so, as the negotiations of the Uruguay Round had been shrouded in secrecy – the claim that it represented a 'consensual' process came under attack, as did the pretensions that the AoA had inaugurated a new era of fairness for the global food economy. The essential criticism of the AoA was that it actually locked in the massive competitive imbalances in the global food economy which had been built up to that point. Watkins (1996: 253) articulated this initial criticism in basic terms: 'far from ushering in a new era for agricultural trade, the Uruguay Round of GATT marks the latest phase in the emergence of a global food system structured around powerful vested interests based in the North to the detriment of poorer people in the South'. A key part of this criticism was that the new discipline given to agro-subsidies lacked teeth. Because the lion's share of agro-subsidies were concentrated in the USA and the EU to begin with, very uneven levels of 'trade-distorting' subsidies would still remain after reduction commitments were undertaken. Further, given the domestic politics surrounding agriculture in the USA and the EU, it was anticipated that their cuts to the disciplined amber box would largely end up getting shifted into the permissible green box. Related to this criticism was the question of whether such a clear line as had been drawn between trade-distorting and non-trade-distorting subsidies was in fact possible, or whether instead the large disparities in green box supports inevitably contribute at some level to structural overproduction and affect international prices and competitiveness. Thus, in effectively sanctioning much of the US and the EU subsidy regimes while at the same time intensifying liberalization – or, at the very least, reducing future border flexibility with tariff levels vulnerable to deeper cuts in the future – all in the wake of structural adjustment, which had already propelled liberalization and cut agricultural support programmes in most developing countries, critics argued that the AoA exacerbated the problems facing most of the world's non-industrialized, non-subsidized farmers within the global food economy.

The imbalances and extensive authority of the AoA came into focus as governments rewrote national policies, as agricultural trade deficits expanded in many poor countries and as the rules-enforcing Appellate was aggressively used to challenge non-compliance, with the proportion of disputes over agriculture much higher than agriculture's place in overall world trade. Not surprisingly, given how the AoA was written, the USA led the way in forcing compliance through the Appellate, with the most appeals and victories, leading US agricultural officials to boast about how they were using the WTO 'as a tool to break down agricultural trade barriers' to an extent that 'the mere threat of US action in the WTO has helped to open markets for American agriculture'. By the time the WTO was approaching the Seattle Ministerial in 1999, USDA and other government officials were calling the AoA a 'landmark' victory and celebrating how 'aggressive trade policies' coupled with 'careful, persistent' efforts 'promoting US products [and] penetrating new markets' had 'paid off handsomely', as 'global demand surged and US exports exploded'.[6] Though access to the Appellate was equal in theory, in practice impoverished governments had a hard time reconciling and affording such recourse given the considerable resources needed for the highly technical research, monitoring and litigation.

While the AoA was the most contentious aspect of the WTO's regulatory dominion, another chief criticism that emerged soon after the Uruguay Round surrounded the TRIPS protocol. Critics challenged the logical framework of the need to secure technological rents in order to provide investment incentives in a variety of ways, attacking the legal biases in how intellectual property gets defined (for instance, is seed selection conducted over many generations considered intellectual property? Or does it become intellectual property when a corporation specifies, manipulates and patents its genetic code?), stressing the immeasurable global disparities in scientific and legal resources (over 90 per cent of internationally sanctioned patents are held by rich-country-based TNCs) and disputing what motive force, the right to profit versus democratically determined social and environmental concerns, should be governing how innovations are put to use. One prominent counter-narrative was that of 'biopiracy', which describes the process of corporations collecting and patenting genetic resources such as seeds and traditional medicines, especially from the biodiverse tropics, which often rest upon the long-term plant breeding and ecological knowledge of local people – effectively 'pirating' as their own intellectual property what was the common property of societies and cultures. The result is that the wealth derived from the genetic resources of the tropics is transferred from mostly poor countries to wealthy corporations. Another notable way in which widening the reach of US-styled patent law would soon begin to play out was in the resistance of pharmaceutical TNCs to the production and distribution of generic medicines to millions of the world's poorest people who could not otherwise afford their monopoly prices.

At the same time as the imbalances in the WTO were being unpacked by government bureaucrats, scholars and activists in the early implementation phase following the Uruguay Round, the USA, the EU and Japan sought to expand its mandate at the Singapore Ministerial in 1996. In addition to pressing for further liberalization in existing realms such as agriculture there was also a move to bring new disciplines to government policies on investment, competition, trade facilitation and the procurement of state contracts, which together became dubbed the 'Singapore Issues'. Developing country negotiators refused to discuss these issues, however. While this was an important moment, slowing down if not yet reversing the momentum of the WTO, it was interpreted in very different ways: for the governments of the world's most powerful countries, their corporate lobbyists and other advocates of the WTO it was merely a speed bump; for developing countries and the growing critics of the WTO it was an indication that the extension of the institution was not a foregone conclusion. This divergence would soon play out much more dramatically in the Seattle Ministerial in 1999, by which point the scope and imbalances of the WTO had come into sharper focus.

There were many areas of disagreement in Seattle, prominent among which were the continuing pressure and resistance to the Singapore Issues, expanding the GATS, deepening TRIPS protections and cutting industrial tariffs, each of which in a general sense had a developed- versus developing-world character, with the latter not only resisting a more expansive WTO but hoping to revisit and rework some of the original tenets while growing increasingly impatient with the lack of transparency and the back-room power politics of the negotiations. Amid the various tensions, however, nothing was more conflictive than agriculture. Debates over the AoA also partly played out with developed and developing countries in opposing camps, but there were other complex layers as well. One complaint was that in the process of 'tariffying' non-tariff barriers some developed countries had established protectionist 'tariff peaks' on the processed goods on which developing countries held a comparative advantage, which were inhibiting diversification into higher-value stages of commodity chains by preventing competitive access to these lucrative markets. A bigger source of contention centred on the discrepancy between the promise of subsidy discipline, which was so integral to the logic of liberalization, and the reality that aggregate levels of state support for agriculture had actually increased in the USA and the EU from the late 1980s to the late 1990s with the early concern of critics borne out as subsidies were shifted from one box to another. According to the World Bank (2002), total agro-subsidies averaged 1.3 per cent of GDP in OECD nations in 2000 or roughly US$1 billion per day – an estimate that exceeded direct expenditure by factoring in the estimated market distortions that these subsidies cause, but which was conceptually striking and oft repeated. For the countries of the Cairns Groups their criticism centred on how these subsidies distorted the competitive playing field and unfairly impeded their export growth. Many non-Cairns developing countries were similarly concerned with the unevenness of competition and drew attention to how import pressures from the agro-export giants threatened to have a disastrous social impact in many places, though as will be seen this was not a straightforward argument for food-deficit LDCs.

While the USA and the EU continued trading barbs over one another's subsidies, by Seattle a more serious agricultural rift had also emerged between them: the issue of GMOs. With the revolving door between the USDA and large agro-TNCs also manifested in US negotiating teams from the Uruguay Round onward, the USA has been a staunch advocate at the WTO of the 'substantial equivalence' case for GMOs laid out by agro-input TNCs like Monsanto. As discussed in Chapter 2, this involves the claim that GMOs are close enough to non-GMO food crops to pose no significant risk to human health or the environment but unique enough to be considered intellectual property. Where this is the legal-scientific norm, as in the USA, there are lax and short-term standards for proving the food safety and environmentally benign quality of new crop varieties, and no legal grounds to force the labelling of GMO products so as to allow consumer choice. This position, if projected multilaterally, would impede the ability of national governments everywhere to establish their own regulatory systems for new crop varieties and use technical barriers to prohibit GMO seeds, food products or feed stock on either environmental or health grounds, because to do so could be classified as a 'discriminatory' trade practice, making it likely to be challenged then punished through the WTO Appellate. This position was justified with the claim, repeated from agro-input TNCs, that resisting GMOs was actually impeding the corporate crusade against world hunger. In sharp contrast the EU wanted the multilateral legal norm to be the 'precautionary principle', which would provide the grounds for governments to enact robust legislation to restrict the use of GMO seeds and to require strict labelling and traceability on processed food and feed derivatives. If upheld multilaterally – consistent with the Cartagena Protocol of the Convention on Biological Diversity (1992), which the USA notably did not sign – this would imply that governments maintain the right to block the importation of GMOs and to set their own standards for testing new crop varieties, food labelling and traceability (Rosset 2006; Buttel and Hirata 2003). As will be discussed in Chapter 5, this is one very important front that highlights the possibilities of education and activism in affecting macro-scale debates, because while the EU is home to some of the world's largest agro-TNCs, popular education, activism and the widespread concern over the issue there have forced European governments to take firm positions against GM seeds and the import of GMO crops.

This also connects to the second dimension of the 'Battle of Seattle', the famous eruption of a new globally networked activism. In Seattle, waves of protesters challenged the WTO as both a symbol and a key instrument in corporate-driven globalization, with the pageant of resistance including human barricades that physically kept negotiators from meetings, property assaults on targets such as Starbucks and McDonald's, and teach-ins, street theatre and other celebrations to mark the hope that 'another world is possible'. The activism on the streets of Seattle had a symbiotic relationship with the opposition by developing countries in the official negotiations, and together global attention was shifted away from further liberalization and towards a more critical view of the WTO. And with the all-or-nothing character of Ministerials the result was gridlock, climaxing with many developing-country negotiators walking out on the last day. For agriculture specifically, Seattle helped to bring the regulation of agricultural trade out of the shadows of legal technicalities and served to identify the WTO much more widely as being critical to the global food economy.

Unlike that of Singapore, the outcome of the Seattle Ministerial could not be written off as a speed bump in expanding the multilateral governance of the global economy. Even its champions understood this as a profound crisis of legitimacy, and intensive lobbying was needed to get the institutional momentum churning again. Discussions were restarted in 2000 with the litmus test being whether a new round of negotiations could be relaunched at the next Ministerial in 2001.

Moving from impasse to impasse

The next Ministerial was, not coincidentally, placed far from the reach of activists in Doha, the capital city of Qatar on the Persian Gulf. At the Doha Ministerial in 2001 the USA and the EU in particular continued to press for expanding the WTO's authority, with the USA playing on fears of a worsening global recession following 9-11, while most developing countries had not moved from their stance in Seattle that the starting point for any negotiations should be remedying the imbalances and implementation failures of the Uruguay Round, and that the process needed to be more transparent. Apart from the official deliberations, activist and scholarly critics struggled with how to strategize opposition to the WTO, with some framing another round of negotiations as a 'damned if you do, damned if you don't' dilemma; that is, if another round were to proceed as per usual it was likely only to intensify US-EU-Japanese demands for liberalization, but if victory meant derailed negotiations, as in Seattle, the grave imbalances of the Uruguay Round would simply be frozen.

Differing positions on agriculture were again front and centre at Doha. The USA and the Cairns Group continued their basic effort to extend agricultural trade liberalization, and the importance that the USA attached to this matter was summed up in a claim by George Bush Jr earlier in the year 'that the cornerstone of good trade policy is good agricultural policy … The job of this administration must be to open up more markets for agricultural produce … and it starts with an administration committed to knocking down barriers to trade.' At the opposite end of the spectrum was the government of India, with an enormous agrarian population to consider. India emerged as the leading campaigner for the case that governments should retain a large measure of policy flexibility with respect to agricultural trade liberalization on the grounds that agriculture has a much broader range of functions in societies than other economic sectors, such as its role in food security, at the root of cultures and in the stability of rural areas, all of which cannot be reduced to an economic yardstick.

In practice, the 'multifunctionality' argument for policy flexibility was associated with demands for two key policy tools together known as 'special safeguards', with many assuming that access would be skewed towards developing countries, though the argument initially came from the EU. The first was that governments be allowed to identify 'Special Products' which would be immune from liberalization commitments (without risk of challenge or penalty) based on their own food security, environmental, cultural or rural development justifications rather than governed by a singular logic. The second was that governments could use 'Special Safeguard Mechanisms' (SSMs), which essentially meant the right to raise tariff levels whenever an import surge in some area was found to have a negative impact on domestic production, also with the locus of agency at the national level irrespective of WTO formulas and procedures. Meanwhile, with this wide gulf over liberalization the criticism about subsidies from both the Cairns Group and some non-Cairns Group developing countries was beginning to sharpen into more aggressive demands to amend the subsidy box categorization and ramp up their disciplines. In the face of this criticism and preoccupied with getting the negotiations moving again, the USA and the EU worked more concertedly together, which meant setting aside their conflict over GMOs for the time being, as well as disputes in other sectors.

In the end, after an all-night meeting on the last day compromises were made, including vague promises to developing countries, and a new set of WTO negotiations was not only launched but heroically branded as the 'Development Round', with a completion date set for 2005. A US negotiator went so far as to insist that at Doha the WTO had 'removed the stain of Seattle'. One important compromise was the concession by developing countries to discuss the expansion of the WTO's authority into new realms, which they had resisted since Singapore, in exchange for a promise that their development needs and the negative fallout from liberalization commitments would receive serious consideration, particularly with respect to agriculture. On this point, the 'multifunctional' role of agriculture was acknowledged, though what this would entail was unclear. Another vague but significant promise to developing countries was that their public health needs would be put above intellectual property rights so that the production of cheaper generics could make desperately needed medicines more accessible to the world's poor. But Doha was also characterized by high tensions, the power politics of exclusionary meetings and carrot-and-stick lobbying by the USA and the EU, such as dangling foreign aid or threatening to revoke it (Jawara and Kwa 2003). One clear manifestation of the recurring imbalance was the EU's success in watering down the initial draft declaration which had made the phasing out of agro-export subsidies an explicit goal for the negotiations (getting the wording 'without prejudicing the outcome of the negotiations' added, in effect eliminating its essentiality to the all-or-nothing negotiations). Thus, many left far from convinced about the sincerity of the 'development' moniker given to the Doha Round.

So it was not surprising that the Doha Round quickly degenerated into another impasse. Again, there were various reasons for this beyond the AoA, but agriculture was at the forefront. The complaint about selective protectionism recurred, backed by UNCTAD studies which showed systemic biases in the deployment of trade measures against processed and semi-processed developing-world exports. But the biggest lightning rod was again the matter of the USA and the EU agro-subsidies, especially as they would not make substantive commitments for reductions while continuing to exert pressure for more liberalization. Exasperation over the subsidy issue grew further with the 2002 US Farm Bill (formally called the Farm Security and Rural Investment Act), which guaranteed an additional US$180 billion in agro-subsidies over ten years – upping subsidies to record levels, maintaining supports focused on massive-scale corporate farmers, adding a US$10 billion subsidized export credit programme and recoupling some support to production. This was roundly seen as a blatant assault on the professed spirit of the 'Development Round', with Mittal (2003) describing it as 'Robin Hood in reverse', effectively 'robbing the world's poor to enrich American agribusiness'. The EU even feigned outrage while, under pressure from France, shirking reductions to CAP spending and downplaying selective protectionism. The 2003 CAP reform was largely an exercise in shifting farm supports into categories classed as 'non-trade-distorting', cutting price supports (to the satisfaction of Europe's agro-TNCs) and direct export subsidies while increasing direct income payments to farmers. Once implemented, it was estimated that the overwhelming majority of the CAP would fall into the green box.

In the face of this criticism, the USA attempted to use its agro-

An uneven playing field

subsidy levels as negotiating leverage, seen most clearly in the insistence by the chief US trade negotiator Robert Zoellick that the US position was strengthened rather than weakened by the 2002 US Farm Bill (Becker 2002). Such posturing did not sit well with the Cairns Group, with countries like Brazil, Argentina and Australia increasing their calls for a less distorted exporting playing field and enhanced access to US and EU markets. Non-Cairns developing countries also continued to feel marginalized, while emphasizing the 'multifunctionality' case for their right to maintain flexibility at their borders through special safeguards. The case for maintaining tariff flexibility in perpetuity was weakened, however, by some more specific arguments for countervailing mechanisms to be set relative to rich-country subsidy levels and for linking future tariff cuts to subsidy reductions. Given the fuzziness surrounding what is and what is not a 'trade-distorting subsidy' and difficulties in measuring the impact, a position that tied liberalization to subsidy reduction commitments effectively legitimized both Zoellick's claim that bigger subsidies make for a bigger bargaining chip and the idea that while subsidies might interfere with the goal of further liberalization the goal itself was beyond contestation, a slippery slope that is examined further near the end of the chapter.

The extent to which the subsidy issue in particular had been magnified as an obstacle in WTO negotiations was evident in the fact that WTO and World Bank officials began to speak out on it. In 2002, WTO president Michael Moore insisted that significant reduction of the agro-subsidies in developed countries 'would return more than five times all the development assistance to developing countries', and World Bank president James Wolfenson argued that 'rich countries must take action to cut agricultural subsidies that rob poor countries of markets for their products', noting how 'agriculture subsidies constitute a heavy burden on the citizens of developing countries' while mainly benefiting large corporations. In 2003, the World Bank's chief economist, Nicholas Stern, called rich-country agro-subsidies a 'sin … on a very big scale', and within the USA the *New York Times*, a flagship of the corporate media, was consistently criticizing the imbalances associated with the agro-subsidy regimes; one editorial called the system 'morally depraved' and the 2002 US Farm Bill 'outrageous', and characterized these as 'harvesting poverty around the world' (NYT 2003). Activists and scholars sought to draw attention to this 'moral depravity' in various ways, highlighting how: the total OECD farm subsidies in 2001 were more than six times greater than its total development assistance to the developing world, with twelve days' worth of these subsidies equivalent to the total annual OECD development assistance for sub-Saharan Africa; the average EU subsidy per cow was roughly the same (US$2) as the poverty level on which billions barely subsist; and the USA was giving more than US$3 billion in cotton subsidies to roughly 25,000 large-scale farmers, which was estimated to depress the world market price from the mid-1990s onwards by 25 to 50 per cent, with devastating impacts on the 11 million farmers in West Africa for whom cotton is their main source of income. Beyond agriculture the noble branding of the 'Development Round' was further deflated by the generic drug issue, as promises of improved access to generics for the world's poorest were turned on their head by the advocacy of the Bush administration (which was heavily endowed by Big Pharma) for stronger intellectual property rights commitments.

In the acrimonious lead-up to the next WTO Ministerial in Cancún in 2003 various attempts at forging some sort of pre-meeting consensus all flopped. In agriculture, the draft negotiating document – which called for aggressive action on liberalization and export subsidy disciplines, some action on other 'trade-distorting' subsidies (without revisiting this definition) and modest flexibility surrounding special safeguards – was thrown out. In its place, the USA and the EU again joined forces to issue a new 'framework for negotiations' which reasserted the claim that the bulk of their subsidies should be exempt from reduction commitments because they were 'non-trade-distorting', offered only limited cuts to those subsidies classed as the most trade-distorting and sought to limit special safeguards. But despite loud objections, including a coordinated rebuttal from Brazil, India and China criticizing the impact of subsidy imbalances on world markets, the WTO Secretariat released an official draft text prior to Cancún largely drawn from the US–EU framework, generating resentment not only for its content but for a process that reeked of another back-room deal. As a consequence, as Bello (2003a: 1) noted, 'not even the most optimistic developing country came to Cancún expecting some concessions from the big rich countries in the interests of development', but rather came 'with a defensive stance. The big challenge was not that of forging a historic "New Deal" but one of preventing the US and the EU from imposing new demands on the developing countries while escaping any multilateral disciplines on their trade regimes.'

On a more constructive note, the joint response prepared by Brazil, India and China foretold of the possibility that there might be increased South–South dialogue on trade issues, which in the case of agriculture had previously been inhibited by the Cairns Group divide and from a broader perspective had been mostly dormant since the great backlash of debt and structural adjustment, as discussed in Chapter 3. Political change in Brazil, a leading Cairns member, was an important factor facilitating this. While the Workers' Party (WP) was hardly committed to radical social change, the fact that it had been elected on the strength of its populist appeal to the urban and rural poor meant that at the very least Brazil's trade policy was no longer the sole domain of the country's long-dominant exporting classes. South–South dialogue prior to Cancún was also cultivated by preparatory technical discussions with research and advocacy organizations such as the Third World Network and Focus on the Global South.

Once the Cancún Ministerial began the impasse developed into another Seattle-like confrontation, with developing-country negotiating contingents similarly emboldened by the large activist presence and also this time more organized into new alliances, the most significant of which was the Group of 20+ (G-20+) led by Brazil, India, China and South Africa (so named because its membership hovered around twenty countries at and beyond Cancún). The centrality of agriculture was even more stark at Cancún than it had been before. On the activist front this was marked by the large presence of peasant movements and the stunning self-sacrifice of Korean farmer Lee Kyung Hae, who stabbed himself through the heart bearing a sign that read 'WTO Kills Farmers'.[7] On the negotiating front, the G-20+ consistently focused on agriculture and stressed that most of the world's farming population was being hurt by the current system, while pointing to the fact that they constituted three-fifths of the world's population and an even greater share of its farmers. The three major agricultural priorities identified by the G-20+ were: increased market access, especially to selectively protected markets in rich countries; the reduction of US and EU subsidies; and the need for developing countries to maintain mechanisms to protect their farmers. This combination contained some serious and unresolved tensions, most notably the divergence between enhanced market access (still most clearly represented by Brazil) versus border flexibility (especially represented by India), but it was significant that the most influential developing countries were discussing mutual interests and at least vocalizing the concerns of their small farmers more loudly.

In addition to the G-20+ there were two other developing-country alliances, the G-33 and the G-90, with some similar and some divergent interests from the G-20+ but in a general sense serving to brace the rising voice of developing countries in negotiations. The agricultural focus of the G-33 was on promoting the new idea of a 'development box' that would be available to poor countries and include secure access to a set of policy tools to protect markets in key areas, centred on their rights to special safeguards. The G-90, whose membership also drew from the above two groupings, was largely composed of ACP members long focused on shared trade relations with Europe, non-ACP members of the African Union and other LDCs. ACP nations were in a difficult, somewhat ambiguous position. On one hand, they supported the 'development box' concept of the G-33 and the rights to self-designated flexibility, as well as the G-20+ calls for increased agro-subsidy disciplines and reducing selective protectionism against value-added processing in developed-country markets. But on the other hand, countries in the ACP wanted to maintain their preferential trade agreement with the EU, which was seen as a historical obligation, and their position was further complicated by the fact that their earnings from some key exports (principally sugar) had long been inflated not only by protected markets but by how domestic EU subsidies affected prices.

In addition to agriculture, developing countries continued to find sturdy common ground in resisting the Singapore issues, which the EU and Japan refused to give up on, and the US effort to extend intellectual property rights. The general point was that if this round was really about 'development' then the onus was on developed countries to make the most of the concessions, not only because of much older historical issues but because of the more recent history of the Uruguay Round. Such resistance also met a weakened joint resolve from the USA and the EU, which had been somewhat damaged by political (e.g. Iraq) and economic (e.g. debates over GMOs; export tax policies; steel tariffs; aircraft subsidies) differences as well as the respective attention being given to bilateral and regional trade pacts. In short, the tensions of the WTO had arguably never been stretched further, and when the Cancún Ministerial ended up collapsing amid waves of protests and defiant proclamations from newly allied developing-world negotiators, not only was the Doha timetable for 2005 derailed but the entire round of negotiations was jeopardized.

Opponents of the WTO promptly hailed this as a more emphatic 'Seattle', with some hoping it might herald a new era of South–South cooperation and reshaped international relations in multilateral forums that could ultimately lead to a new regulatory framework for trade. Such big hopes were fanned by Celso Amorin, Brazil's trade minister and lead negotiator at Cancún and the unofficial spokesman for the G-20+, who insisted that the strength and resilience of this new alliance were rooted in 'the support of our producing classes and of world opinion in general'. At the end of the meetings Amorin boldly declared:

> We stand united, we will remain united. We sincerely hope that others will hear our message and, instead of confronting us or trying to divide us, will join forces in our endeavour to inject new life into the multilateral trading system. To bring it closer to the needs and aspirations of those who have been at its margins – indeed the vast majority – those who have not had the chance to reap the fruit of their toils. It is high time to change this reality. This should be the spirit of Cancún.

Beyond Cancún: which way forward?

Cancún loudly echoed Seattle's earlier statement that a few dominant governments representing the most powerful economic interests within them could no longer write the rules for the global economy and ram them through as consensus. But beyond exuberant declarations such as that made by Amorin, an essential question that went unspoken was how injecting 'new life into the multilateral trading system' would help 'the vast majority' of heretofore marginalized populations. This question had both practical and strategic dimensions. Practically, the hope that strengthened South–South alliances could move from an oppositional stance and begin to construct a more progressive framework for global economic governance, or even for regional integration, had to be tempered by the fact that there was great ideological diversity and some highly autocratic regimes among allied governments. The fact that many were suddenly rallying around moral appeals to their 'producing classes' in opposition was not the same as being prepared to engage in difficult political initiatives on their behalf, which Amorin's WP government was demonstrating at home with their lack of action on land reform. The practical challenge for new South–South alliances building on Cancún also involved facing up to the serious differences in priorities which the rhetoric of solidarity had not resolved: countries with competitive agro-export platforms such as Brazil, Argentina and Thailand maintained their calls for aggressive liberalization and disciplining rich country subsidies; large countries facing dramatic social convulsions such as India and China were most concerned with maximizing their flexibility in how they engaged with the global food economy; the ACP wanted to hold on to its trade preferences with the EU or at least receive some compensation; and the poorest, net-food-importing countries dependent upon cheap surpluses from the temperate world viewed increased subsidy disciplines with more caution given that this could, at least in the short term, lead to more expensive imports. The case of sugar soon brought some divergent interests to a head, because at the same time as Brazil was speaking out about the need for SDT measures for small and poor economies it was preparing its case along with Thailand and Australia to challenge the EU–ACP sugar regime at the WTO; subsequently adjudication went against the EU–ACP in 2005 (the EU–ACP had already previously lost a similar case with bananas which had been brought to the WTO by the USA and five Latin American banana exporters at the urging of Chiquita).

Another practical obstacle to the new developing country alliances was that with multilateral discussions stalled it was expected that the USA, the EU and Japan would increasingly pursue bilateral and sub-regional pacts with attention centred in Asia. Chief US trade negotiator Robert Zoellick, in fact, posed rising US bilateralism as a direct threat after Cancún, touting such things as the Central American Free Trade Agreement which the USA was in the midst of negotiating. This in turn entailed the possibility of the USA and the EU using their big markets, sources of investment and foreign aid as leverage to 'divide and rule'. The attendant risk was that if action on institutionalizing trade rules were to become concentrated at bilateral and sub-regional levels the arm-twisting and power politics so familiar to multilateral negotiations would be magnified further, such that developing countries could end up coerced into greater concessions than would be possible at the WTO on such things as tariffs, intellectual property rights and protections for investment and profit repatriation.

The strategic question about how the multilateral trading system could be reinvigorated so as to help 'the vast majority' was essentially about whether the struggle for alternatives should be inside or outside the WTO. In agriculture, for those intent on challenging the course of the global food economy this meant debating to what extent binding multilateral agricultural trade regulation was necessary and whether the WTO could be renovated to serve the interests of small farmers or whether the fight should instead aim to get agriculture out of the WTO – or tear down the institution altogether – and rebuild alternative forms of economic regulation through different forums and at different levels. The strategic question about whether the objective was to *renovate* or to *raze* the AoA and the WTO was a divisive one. In agriculture, some small-farmer organizations, activists and progressive academics were quick to claim that the events of Cancún represented a victory for the world's small farmers and a step towards razing the AoA and perhaps even the entire WTO, but others felt that while the AoA and WTO were badly flawed there remained a need for a multilateral regulatory system for the global food economy.

The 'need to renovate' position was a revised version of the 'damned if you do, damned if you don't' attitude noted earlier, which saw WTO negotiations as being unavoidable, with the significant amendment that the negotiating playing field was thought to have become much more level with the emergence of the South–South alliances. There was a range of ideas about what these renovations might target, with some so sanguine as to suggest that WTO rules could be fundamentally reworked to incorporate environmental priorities or United Nations commitments on human rights. At the other end of the spectrum were those who accepted either the logic or the inevitability of further market integration and sought to focus renovations not on the WTO's foundational principles but on their uneven application – a view often accompanied by the charge of 'hypocrisy' for the way developed countries preached the merits of liberalization, craftily invoking it when it suited exporting interests but keeping significant protections in place when it threatened certain domestic industries. From this vantage point, renegotiating the WTO was now 'damned if you don't, and maybe rewarded if you do', a position that is encapsulated well by Oxfam's Watkins (2003: 7):

> Some anti-globalizers will view any proposal to reform the WTO as ill-conceived. But what are the alternatives? If you want a glimpse into the future of a world with a weakened multilateral system take a look at the content of regional and bilateral trade pacts … The WTO's rules are rigged in favour of the strong. Yet abolition is not an option. Apart from removing a source of pressure on the US and the EU to open markets, cut farm subsidies and halt protectionist abuses, it would risk a ruinous spiral of conflict. Rich countries would bulldoze poor ones into deeply unequal trade treaties.

Watkins (ibid.: 2) insisted that recourse to '"people-based" or regional alternatives, allied to a withdrawal from the global economy in favour of localized exchange' rests on a flawed assumption 'that the power politics that distort trade relations between rich and poor countries will wither away' with the demise of the WTO. Such an account, he argued, 'combines implausible politics with silly economics. Lacking the means to retaliate poor countries need rules and a multilateral system that works. Trade provides one avenue through which poor countries can tap into more prosperous markets, and gain access to new technologies and ideas.' Other keen critics of the WTO expressed similar views after Cancún, with Das (2003a) having called upon 'all concerned with international trade' to 'work for salvaging the WTO from the debris', and Khor (2003) claiming that it was 'urgent that measures be taken to turn the WTO into an organization that truly respects the developing countries and their development objectives'. These sentiments were echoed by leaders in the G-20+, who quickly began urging the developed countries not to give up on multilateral negotiations through the WTO.

In contrast, the 'need to raze' position held that the WTO was irredeemable and that the goal of renovating it was a distracting chimera for anyone struggling for a more socially just world, when the objective should instead be to raze the institution and get on with other efforts to localize and democratize decision-making authority for trade and development policies and, where it supports these goals, enhanced regional integration. Bello (2003b: 2) argued that the WTO should not be seen as 'a neutral set of rules, procedures, and institutions that can be used defensively to protect the interests of weaker players. The rules themselves … institutionalize the current system of global economic inequality.' For Bello, who engaged in a spirited debated with Oxfam over their 'renovate' stance on the WTO,[8] the suggestion that the option is between this or else having uneven North–South bilateral trade deals proliferate is a 'false choice' that draws attention away from work towards 'real alternative arrangements, such as creating regional economic blocs or restructuring economic existing ones such as MERCOSUR and ASEAN to serve as effective engines of coordinated economic progress via policies that effectively subordinate trade to development' (Bello 2003c: 4). In addition, the suggestion that the multilateralism of the WTO provided a defence against proliferating (and potentially more exploitative) bilateralism ignored the fact that there was actually a significant rise in bilateral agreements *after* the onset of the WTO in 1995.

In agriculture, Bello's arguments reverberate loudly in the position of the growing world peasant federation, La Vía Campesina:

> Rather than restricting its efforts at reforming the WTO by negotiating what can be placed in the 'green' or 'amber' box, or making slight adjustments with the creation of a 'development' or 'food security' box, La Vía Campesina argues that agriculture and food should simply be taken out of the WTO. (Desmarais 2002: 105)

The core of this position stems from the fact that most small farmers in the developing world, who comprise the overwhelming proportion of the world's farming population, are entirely focused on domestic markets while it is large-scale farmers who dominate agro-export sectors everywhere. For domestically oriented small farmers, the institutionalization of market integration promises only increased exposure to uneven competitive pressures from industrialized systems, while the gains from enhanced access to developed county markets accrue to the elite minority of the farming population in developing countries. But because most governments of developing countries are, as Rosset (2006: 82) notes, 'beholden to small but immensely powerful agro-export elites', they 'have bitten hard on the market access bait' as a primary negotiating objective. Meanwhile the uncertain prospect that future multilateral negotiations might achieve greater disciplines for developed-country agro-subsidies does not make the prospect of further liberalization much more palatable for small farmers. Subsidy disciplines or not, vast scale and technological disparities and the massive externalization of environmental costs would still allow industrialized farmers to operate with much lower margins than do smaller farmers, and aggressive corporate marketing would continue to drive dietary shifts. Thus, La Vía Campesina has argued that if the rules for the global food economy were written to serve the interests of small farmers they would prioritize 'food sovereignty' and the policy flexibility of governments, and not breaking downing barriers to export markets (Rosset 2006; McMichael 2004b; Desmarais 2002).

This interpretation helps to make sense of the flipside of the recurring WTO conflicts: its resilience. While North–South imbalances were the most prominent and publicized aspect of the politics of the WTO, differential class interests *within* developing countries – namely the priorities of trading and producing elites – were a big part of why these governments kept returning to the negotiating table and why, less than a year after Cancún, the Doha Round was officially restarted at the WTO headquarters in Geneva.

The uncertain future of the WTO

In July 2004, hidden from the global spotlight and from most WTO member states, negotiators from the USA, the EU, Australia, Brazil and India (the latter two selected to represent the G-20+) announced that they had put the Doha Round 'back on track' by trading off US and EU promises for more subsidy disciplines in exchange for developing countries committing to negotiate further liberalization in agriculture and other sectors, including a tariff-cutting formula featuring more aggressive cuts to higher tariffs than to lower ones. How such highly exclusionary meetings and the so-called 'Five Interested Parties' could restart negotiations out of the previous impasse and set the framework for the next Ministerial was somewhat mystifying, and any pretence that the WTO may have still been clinging to at this stage about being a democratic forum of nations was unquestionably severed here.

Each party emerged claiming a victory for their interests, with Brazil and India both framing their gains in ways that made Cancún's promises to reshape trade regulation in the interests of their working people seem like a distant memory, focusing on how the rules were skewed against developing-country exporters. Brazilian negotiators focused on assurances (though no concrete commitments) that direct export subsidies would be disciplined, something of keen interest to Brazil's exporters but unimportant to small farmers since Brazil is not a significant food importer. Indian negotiators trumpeted a few assurances that were of interest to high-tech and service sectors while failing to get a promise that special safeguards for agriculture would be secured, for which they were slammed in their national media for sacrificing India's farmers to its new economy. But while the peculiar Geneva restart officially breathed new life into the Doha Round it didn't resolve the tensions that continued to simmer into the Hong Kong Ministerial in 2005.

Though the Hong Kong Ministerial did not draw the same global attention as did Seattle or Cancún, negotiations were similarly combative. The G-20+ was again a formidable negotiating force, despite serious divisions surrounding special safeguards between agro-export powerhouses like Thailand and Argentina, which wanted these framed restrictively, and countries like India, Indonesia and the Philippines, which wanted these framed more liberally. India in particular re-emerged as the leading defender of the position that developing-country governments should retain substantial flexibility with respect to protecting domestic agricultural markets, pointing to its massive farming population and food security concerns and making it clear this time that it was not prepared to sacrifice these with further liberalization commitments. Indian officials also stressed the early promises that this round of negotiations was supposed to be about 'development' in the South, and hence focused on correcting original imbalances rather than prising open developing world markets any further. Meanwhile, other developing-country blocs continued to line up with proposals at odds with those of the USA and the EU, also with the idea that they were not interested in 'horse trading' and that there was instead an uneven burden of responsibility in making concessions. This position was fortified by World Bank and UN agency studies which projected that the plans for liberalization set out in the negotiating framework would bring much greater gains to the developed world than they would to most developing countries, and that within the developing world gains would be concentrated in a small number of countries while the poorest countries would be the most adversely affected.

Again the result was an impasse, though unlike in Cancún negotiations were not stalled entirely but kept limping along. The slim continuance of the Doha Round involved concessions by developing countries on agricultural market access in exchange for more vague promises about the USA and the EU disciplining subsidies within the coming decade. But a series of impediments soon reared up once more, principally: the lack of real movement by the USA on disciplining 'trade-distorting' domestic subsidies (without touching the definition of trade distorting) while criticizing special safeguards and calling for greater market access; the EU reluctance to reduce its selective market barriers in agriculture; the unwillingness of large developing countries to reduce their industrial and agricultural tariffs and open markets for services; and the recurring US–EU conflict over GMOs (with the US chief negotiator for agriculture having formerly headed the American Seed Trade Association and worked as a high-ranking executive with GMO company Dekalb). The latter conflict was marked by the US filing a complaint with the WTO aimed at weakening the ability of governments to legislate with respect to GMOs, which was nothing less than a legal attempt to trump widely held scientific and ethical objections. The lack of movement on market access caused the USA to largely lose interest in the round and set its sights on bilateral and sub-regional agreements where it felt more capable of securing its interests. In July 2006, with the stage shifted back to Geneva and Japan added to the roster of the 'Five Interested Parties' (making it now the G-6), there was a different outcome to that of two years earlier. Following more gridlock on agro-subsidies and agricultural and industrial tariffs, Pascal Lamy – the former chief negotiator for the EU who had become the Director General of the WTO – announced the official suspension of the Doha Round, stating that: 'There's no beating around the bush. We're in dire straits.'

It is unclear whether Lamy and other key actors expected the suspension to really just be a short-term 'time out' for negotiations, serving as a dramatic pause for thought, or rather whether they foresaw it as the onset of a longer-term hibernation. *The Economist* (2006a: 11), an ardent supporter of trade liberalization, predicted that not only was the 'wrecking of the Doha round' likely to be 'terminal', but that 'in the long run, the lack of commitment to multilateral trade that sank the Doha round will also start to corrode the trading system as a whole', a forecast that led it to describe this as 'the biggest threat yet to the post-war trading system'. Another noteworthy interpretation came from Brazil's Celso Amorin, one of the loudest voices of opposition in Cancún and subsequently of the 'need to renovate' camp, who called the suspension 'a disappointment' and described the WTO as 'irreplaceable'. Oxfam, also an outspoken voice for reform, similarly identified this as a 'tragic outcome'. This sense of 'disappointment' and 'tragedy' about the WTO's hibernation was reminiscent of the 'damned if you don't' attitude towards negotiations – that is, even with negotiations in their current purgatory the WTO continues operating – and in terms of agriculture this reflected the importance some vested in increasing the disciplines on US and EU agro-subsidies and selective protectionism. This position also stemmed from an apprehensive view of rising bilateral and regional trade agreements as being more apt to be skewed against the interests of smaller and poorer countries.

But it did not take long for this tossing and turning in the WTO's hibernation to transform back into yet another negotiating restart. In February 2007, roughly half a year since he had described the WTO as being in 'dire straits', Lamy stated boldly that 'political conditions are now more favourable for the conclusion of the talks than they have been for a long time. Political leaders around the world clearly want us to get fully back to business, although we in turn need their continuing commitment.' Thus, regardless of whether or not one believes that the WTO can be reformed to better serve the needs of the poor majority within developing countries, for the indefinite future it seems likely to remain at the forefront of debates over how to reorient the trajectory of the global food economy.

Conclusions

In attempting to make sense of the major forces that established and have sought to extend the WTO there is, on the one hand, a need to follow the national governments that are the ones doing the negotiating and signing the deals and, on the other, a danger that this can obscure the fact that the WTO is first and foremost an institution designed to serve the interests of transnational capital. Agro-TNCs have been able to shape the multilateral negotiating agendas and rules-making through their very successful exertion of influence upon the governments of countries where they are most heavily based, with the state–capital nexus deepest and most influential in the USA, which

The Economist (2006a: 11) euphemistically describes as the 'animating spirit behind earlier trade rounds'. In agriculture, one vivid example of this influence has been in how the systematic alternations between corporate executives and senior US government officials discussed in Chapter 2 have been manifested in key actors throughout the cycle of negotiations.

In a general sense, the corporate desire to influence multilateral negotiations stems from the fact that a favourable and expansive regulatory framework can help to widen its scope for accumulation. For agro-TNCs in particular the great prize in multilateral trade regulation was to maximize their global flexibility by restricting the capacities of governments to intervene in domestic markets, while also ensuring that governments would uphold stringent intellectual property rights, reflected in the incessant pressure for expanded market access in the AoA and for strengthening the TRIPS protocol. Agro-TNCs are not disinterested in the subsidy regimes in the USA and the EU – indeed, they have been the major beneficiaries along with the largest-scale farmers – but they are most concerned with market access and would happily trade a measure of subsidy discipline for liberalization – with the important provision that this subsidy discipline does not involve their great anathema, supply management (Rosset 2006).

Agro-food TNCs know that their ability to source cheaply is enhanced when the world's farmers are locked into borderless competition with one another, without bounds on the volume of farm production but with legal bounds on how the nature of farm production can be regulated. If this can be accomplished and a large dose of subsidies left in place to help prop up the biggest, most industrialized farmers facing very low or even negative margins without interfering with the farm-gate prices, then it is all the better for agro-TNCs, as it helps to fortify the system, as do the 'green box' subsidies that support corporate capabilities in multifarious ways. Meanwhile large, industrialized farmers themselves constitute an influential lobby in both the USA and the EU, with an obvious reflection of this being the fact that despite all the wrestling over CAP reform in the EU, in 2006 the CAP still accounted for roughly two-fifths of the total EU budget at the same time as agriculture employed only about 2 per cent of the EU workforce. So with the massive and uneven agro-subsidy regimes continuing to be largely sanctified in the Uruguay Round and through subsequent negotiations, at the same time as market access was projected as the overriding objective, the WTO has served to entrench dislocating competitive pressures facing small farmers around the world.

Developing-country negotiators and South–South blocs threw a spanner in the process of deepening the regulatory framework established in the Uruguay Round, while the rise of civil society has helped to draw the WTO out of the shadows as something that was geared towards magnifying the control of TNCs at the helm of the global economy. But elites and corporate interests in developing countries have also exerted a disproportionate influence on developing-country negotiating agendas, which has helped to draw them back to the table time and again in spite of the many frustrations about how the overt and back-room power politics, exclusionary meetings and coerced consensus have repeatedly played out. Often this has translated into developing country negotiators focusing on calls for increased access to selectively protected markets in developed countries, which enhances the overall corporate vision of the global food economy, and for subsidy disciplines, which in some ways can reinforce the idea of liberalization as the normative logic; that is, reducing flexibility of states at the border would be OK if only the playing field weren't so tilted by 'trade-distorting' subsidies.

At the outset of this chapter it was noted that the WTO's AoA boasts about bringing 'fairer markets for farmers', virtuous-sounding claims that attempt to portray the economic rationale for these rules as being above politics. Such a benign image breaks down, as this chapter has sought to illustrate, when we understand the AoA's rules, the process through which they have been designed and the highly conflictive efforts to extend them; what emerges in its place is a converse picture of the WTO as an intensely politicized struggle to write very partial economic rules. With the high stakes of WTO negotiations set against the backdrop of immense political imbalances, a lack of transparency, the contradictory tensions within South–South alliances and the threat that even worse regional and bilateral trade agreements might proliferate with a failed WTO, opponents of the WTO have faced difficult strategic questions, and much of the research, organizing and networking underpinning this opposition has focused on what is wrong with the institution itself. This opposition has had some notable successes, most fundamentally in breaking down the WTO's early sense of impregnability – as both a symbol and an instrument in the triumphalism of global capitalism – to the point where its fallibility became unmistakable even to its staunchest advocates.

A very uneven set of rules governing agricultural trade remains in place, however, backed by the institutional capacity to adjudicate and exact trade penalties if these rules are not adhered to, and where future negotiations will lead is highly uncertain. What is certain is that the regulation of global agricultural trade must be confronted by small farmers and their organizations, civil servants involved in agriculture, activists, NGOs and concerned scholars, as this will serve to either widen or restrict the space that exists at different levels for efforts to reorient food economies in more socially just and ecologically rational ways, the subject of Chapter 5.

Notes

1 <www.wto.org/english/thewto_e/whatis_e/tif_e/agrm3_e.htm#tariffs>.

2 The discussion of the WTO draws especially from the work of Khor (2006, 2002), Rosset (2006), Clapp (2006), McMichael (2004a, b, 2000a, b), Das (2003b, 1999, 1998), Stiglitz (2003) and Bello (2002a, b). The WTO's myriad technical commitments are readily available on the internet (<www.wto.org>) along with their official rationale. For agriculture, Rosset (2006) provides an excellent starting point for dissecting these more extensively. For the WTO as a whole, the Third World Network (<www.twnside.org.sg/>) provides incisive and current analyses on the technical commitments contained in the WTO, preparatory proposals, ongoing negotiations and the associated political manoeuvrings.

3 This is a reference to the famous claim by its chairman in the middle of the twentieth century that 'what is good for General Motors is good for the rest of America'.

4 The terminology of 'developed countries' is used in this chapter in place of 'industrialized countries' following the official WTO parlance used in differentiating commitments.

5 *Codex Alimentarius* is jointly governed by the FAO and the WHO. See <www.codexalimentarius.net>.

6 The quotations here come from US congressional testimonies in the lead-up to Seattle (Schumacher 1998; Scher 1998; Glickman 1998a). See Chapter 2, note 5.

7 Rosset's (2006) preface provides an eloquent tribute along with Hae's written explanation.

8 This debate is available at: <www.maketradefair.com/en/index.php?file=30052002111119.htm>.

FIVE

The battle for the future of farming

The (resistible) trajectory of the global food economy

This book has sought to illuminate the major contours of the global food economy and the systematic disarticulation of agriculture from ecosystems, communities and even the authority of nation-states, starting from the premise established in Chapter 1 that this trajectory is exacting a tremendous burden on most of the world's farmers, the environment and other species, with a potentially calamitous fallout. It has approached this by examining how key aspects of disparate food economies were forged and became tied into a fast-widening system, and how this system was eventually insulated by a multilateral institution, the WTO, that was largely designed by the governments of a small number of countries in the interests of very powerful corporations.

The global food economy pivots on the industrial grain-livestock complex of the temperate world (Friedmann 2004, 2000, 1993), which assumed this central position in global agricultural trade patterns as this trade took off in the second half of the twentieth century. The growth and industrialization of the grain-livestock complex reconfigured consumption patterns and gave rise to the world's largest agro-TNCs, whose export ambitions were fuelled by government programmes. Most tropical and semi-tropical countries, on the other hand, faced the onset of increasing agricultural trade from the position of having been made dependent upon a small range of commodity exports during the colonial era, with highly inequitable landscapes and labour relations and corroded dynamics of innovation. The problems associated with this dependence were subsequently aggravated by declining terms of trade, transfer pricing and profit remittances by foreign-owned corporations, protectionism that inhibited value-added processing and the roll-back of the state stemming from debt and structural adjustment.

Since the 1970s, agro-TNCs have been the dominant forces transforming the nature of agriculture and integrating markets, horizontally across space and vertically throughout input and commodity chains. In their quest to increase markets and profits, agro-TNCs are relentlessly forging input dependence and standardizing the nature of agricultural production, subjecting soaring farm animal populations to brutalizing treatment, toxifying soils and water and externalizing environmental costs, reshaping dietary aspirations, breaking local bonds between production and consumption, devalorizing labour and replacing it with technology and progressively appropriating control and surplus value from farmers and farm communities. Control and profits are instead centred in the complex and ever more despatialized corporate webs of agro-inputs, processing, distribution and retailing. This has trapped farmers in a rising (input) cost–falling (output) price squeeze, hurting the viability of small farms in both rich and poor countries, with the latter representing the vast majority of the world's farming population. Amid this competitive squeeze, rising production has gone hand in hand with the polarization of industrial farming, as a small number of very large farmers have survived low margins by growing in scale and technology, with this growth assisted by both the uneven subsidy regimes in rich countries and by the indirect subsidization that is implicit in the externalization of environmental costs. The booming productive model in the global food economy is thus the large-scale, industrialized, subsidized and fossil fuel-intensive monoculture on a treadmill of agro-chemicals and fertilizers, coupled to (either directly to or through an intermediary) intensive livestock production, typically of no more than one animal species. The flipside of this in consumption terms is the meatification of diets, or what Cockburn (1996: 27) describes as the 'surge in meat-eating associated with industrial capitalism'.

In short, ever-greater agglomerations of capital are directing the uneven process of market integration, bringing sweeping social and ecological changes in global food production and overriding important concerns about equity, democracy and the locus of sovereignty, the well-being of rural communities, the place of agriculture in the fabric of different cultures, the impacts of farming on ecosystems, the space in the biosphere left for other species and the treatment of domesticated animals. And yet in spite of the weight of these issues, as has been emphasized at various points, the struggle to build alternatives must contend with the fact that this trajectory has been widely 'naturalized'; that is, it appears to many to be a good, inevitable or irresistible evolution, or all of the above. The illusion of inevitability is given strength by the fact that the rapidly expanding corporate webs directing the global food economy are at once undeniably bountiful, increasingly ubiquitous, boldly branded and yet remarkably opaque to most people, while the institutional fortification is hidden from sight and much attention, apart from the wide visibility of a few conflictive episodes, most notably Seattle and Cancún. This illusion also gains force from the theorized benefits of an integrating global market – centred on promises associated with increasing efficiency – that are echoed repeatedly in the corporate media, as well as from the tendency shared by the right and large segments of the left to see the demise of small farming as a historically progressive development, even if it might induce some social pain in the short term.

At the outset of this book, one of the key objectives laid out was to unsettle illusions of inevitability surrounding the global food economy in the hope that this can help to further invigorate the battle for the future of farming, and by now there should be a clear sense of what this battle is up *against*. But, as Foster (1999: 149) suggests, it is crucial that in the face of prevailing structures and myriad crises we guard against:

> ... the notion that history has in some way reached its end or that there are no genuine alternatives. Such beliefs of received ideology, if followed, would guarantee a barbaric, even apocalyptic, outcome. The real future of humanity, as distinct from this non-future, depends on the nature of our social and environmental movements, and ultimately on our willingness to reinvent human history and our social and ecological relations of production.

Social and environmental movements have always been fired not only by a critical appreciation of the existing order but by imagining how productive relations can be reinvented. In this light, the focus of the remainder of the book is twofold. First, it examines some of the basic dimensions of future systems of farming to be battling *for*, a discussion centred on the matter of efficiency and how this actually points to labour-intensive small farms and not to industrial agriculture, before making a case for small farming that hinges on the idea of dignified work. Rather than a blueprint, the intent is to develop an optimistic understanding of how the battle for the future of farming could lead towards more ecologically rational and socially just systems of food production. The second section then explores some of the key levels at which this battle must be waged and won for a new agriculture to take root.

Battling for what?

Agricultural efficiency and the ecological rationality of small farming In a world of 6.6 billion people today and more than 9 billion by 2050, advocates of high-input, industrial agriculture and genetic engineering argue that the world cannot afford *not* to have this model be the productive basis of the global food economy, guided by competition and comparative advantage in a system of liberalized trade. The argument centres on the alleged efficiency gains of industrial agriculture with the implication that the expansion of this model therefore maximizes the potential food supply, with self-evident benefits for a growing human population (more and cheaper food) in a world where hunger is already rampant. The efficiency argument has also been used to frame the expansion of industrial agriculture as an environmental priority. In essence, the green case for industrial agriculture attempts to portray the choice as being between setting on-farm versus off-farm environmental priorities, and equating industrial agriculture with the latter by lionizing a narrowly conceived measure of productivity.

This argument starts from a conceptual binary in which there are 'human' and 'natural' landscapes, classing agriculture as the former. As a human landscape, farmland itself is then exempted from environmental concerns, such as those relating to on-farm biodiversity, nutrient cycling and minimizing external inputs and toxins, and the ultimate environmental good is instead framed through the lens of yield efficiency and the implied benefits this entails off the farm, for natural landscapes. In essence, it is argued, if high-input monocultures can maximize yields, they will reduce the overall land that needs to be devoted to agriculture and in the process maximize the area that can be set aside from human use (i.e. in parks and other forms of protected areas). This is sometimes accompanied by the obvious point that per-farmer output is much greater with industrial methods.

Conversely, the argument follows, if all of the world's present agricultural land were to be made up of low-input, lower-yield small farms there would be less total food produced, with this lower aggregate supply entailing higher food costs and worsened access for poor people around the world. So while low-input farming systems might have been able to sustain biodiversity over millennia at relatively low population densities, to maintain and expand the world's present food supply from diverse small farms is seen to entail devoting more land-space to agriculture, in turn requiring the further expansion of the agricultural frontier that would leave less land for self-organizing ecosystems and other species (Avery and Avery 2003; Avery 1999; Wood 1996; Smith and Forno 1996). Avery and Avery (2003: 135) take this argument farther by linking it to a call for the expansion of the WTO, arguing that two simple ways to 'increase land use efficiency' are to 'liberalize trade of agricultural and forestry products under the WTO' and to reduce 'international food and trade regulations on agricultural biotechnology products' in order 'to ease the development and adoption of this technology by farmers'.

As has already been discussed in Chapter 1, attempting to equate the expansion of industrial agriculture with the moral imperative of improving access to food for the world's poor rests on shaky ground given that there is already enough food, *one and a half times over*, according to the FAO, to adequately feed the world. Midkiff (2004: 3) puts this bluntly: 'Feeding a hungry world has become an integral part of agribusiness speak, and it is nothing less than the Big Lie.' But the matter of agriculture's footprint and productive efficiency is indeed a crucial environmental concern, particularly in the context of the extinction spasm and the continued loss and fragmentation of self-organizing ecosystems. Using agriculture in its most efficient way does harbour the potential to reduce the land-space needed for food production, a conservation issue that is especially pressing since the best arable lands everywhere are already in cultivation or pasture or else have been degraded or urbanized. Ideally, enhancing the efficiency of agricultural land use could also go even farther and support habitat restoration for other species by allowing more marginal farmland to be progressively taken out of production and given over to self-organizing forests, grasslands and wetlands.

At first glance, the environmental case for industrial agriculture might seem compelling, with the objective of meeting food demand from the least total land-space – and hence containing the off-farm footprint of agricultural systems – the most urgent priority, and one that could reasonably trump on-farm goals of enhancing agro-ecosystem cycles and diversity and reducing inputs and toxicity. There are a number of problems, however, with the green case laid out by advocates of industrial agriculture. One of the most basic of these is the fact that agricultural land occupies a huge portion of the world's surface and plays a role for many species that is simply not equitable with urban or industrial land, and therefore does not fit so easily within the contrived 'human' versus 'natural' landscape binary. Further, this binary treats the toxicity of industrial agriculture as though it occurs in a vacuum and does not end up affecting soils, ground and surface water and non-target species, which is wildly misleading. It also ignores the atmospheric burden associated with heavy fossil fuel usage both on the industrial farm and off it. Given that the advocacy of industrial agriculture is closely linked to trade liberalization and global-scale specialization in theory and practice, the rising food miles of the present age are not a separate issue from the nature of production but rather are intimately linked to it, and the unaccounted cost of carbon emissions into the atmosphere is a significant part of the environmental externalities that implicitly subsidize the comparative advantage of cheap industrial foods. While the calculation of food miles can be very complex, especially if it takes inputs into account as it ideally would, systematically reducing the distance that food travels from farm to plate is a basic objective for building more ecologically rational food economies.[1] In a general sense, this is mutually supportive of the goals of working towards more equitable and diverse farm landscapes, unsettling the dominant place of corporations as intermediaries between producers and consumers and securing for farmers a greater share of the wealth derived from agriculture.

Another major conceptual problem relates to the misleading equation of crop yield efficiency with overall productive efficiency, the premise from which off-farm environmental benefits are theorized. The emerging field of agro-ecology is showing how the superior individual crop yields of high-input monocultures are a very incomplete way of understanding efficiency, and that this does not necessarily translate into a more efficient use of land than that of smaller, labour-intensive, low-input multi-cropped farms.[2] Agro-ecologists reconceptualize how agricultural efficiency is understood, expanding the lens for productivity from specific crop yields to net output per unit area while emphasizing that efficiency is also about process and not only about a measurable end point, with small farms typically much better at retaining and recycling organic materials and hence drawing on local, renewable resources. Shifting the measure of productivity from crop yield to net output per unit area and embracing a more holistic view of efficiency for agro-ecosystems strike at the crux of the green case held up for industrial agriculture, because reducing the toxicity and carbon emissions from agriculture while also reducing the volume of land needed for food production come to be understood as intertwined rather than diverging priorities.

Measuring net output per unit area entails aggregating the production from both small, intensive, multi-cropped farms and monocultures over the same total area, and when this is done on similar-quality farmland the total production from small farms has often been found to mirror or even exceed the total of the same crops grown in industrial monocultures, as well as providing greater stability over time. There are a number of reasons why small farms have the capacity for equal or greater net output than industrial farms, and why crop yields and production per farmer are such a skewed way to assess efficiency. As a general rule, labour intensity declines quickly with expanding farm size as labour is replaced by mechanization, and with fossil energy the basis for the greater per-farmer productivity of industrial farms and livestock operations this obviously entails a much bigger carbon footprint per output. Given how the real costs of fossil fuel consumption are unaccounted and hence subsidize both the production and the typically high food miles associated with industrial agriculture, the age of 'peak oil' and rising fuel costs – whether induced 'naturally' in the market by pure demand and supply or 'artificially' by government implementation of carbon taxes in response to climate change – costs could well prove to be one of the most destabilizing forces changing the course of the global food economy, and corroding the competitiveness of industrial foods vis-à-vis local and organic production.

Increased mechanization goes hand in hand with monocultures, which leave bare ground between planted rows. Conversely, small farms tend to make more intensive use of such space by cropping patterns that integrate complementary plant species and where human labour cannot easily be replaced with machinery, as well as integrating small livestock populations and using ecologically benign animal draught. Denser multi-cropped patterns that are frequently rotated, occasionally fallowed and integrated with small livestock and draught animal populations foster decomposition-assisting soil micro-organisms, detritivores and invertebrates, which enhance the biological regulation of soil fertility, pests, weeds and disease cycles with much fewer chemicals and fertilizers than monocultures. This ability to renew and even enhance soils over time is also a function of the radically different time scales of farm management: because small farms are often drawing on knowledge passed down through generations on the land they tend to be organized with a much longer-term objective of equilibrium, in contrast with how industrial monocultures are governed by an annual balance sheet.

The assumption that maximizing crop yields on industrial mono-cultures will tend to reduce agriculture's aggregate footprint and the colonizing pressure exerted is another untenable abstraction. Even if there were definitive efficiency gains to be had on industrial farms rela-tive to smaller, labour-intensive farms, these would not tend to reduce the space agriculture occupies in the landscape. The industrialization of farming, from the temperate grain-livestock complex to the Green Revolution, has greatly exacerbated inequalities in land distribution, but this process at one end does not simply produce urban migration or off-farm rural employment at the other, recalling that the world's absolute rural population has still grown slightly in recent years, though shrinking significantly in relative terms. As a result, the social convulsions associated with de-peasantization are not only linked to 'a planet of slums' but mean that desperate small farmers and landless people will continue pushing agricultural frontiers farther into marginal (e.g. hilly, erosion-prone, arid) areas where soil fertility is difficult to maintain in the long term. From the perspective of biodiversity conservation the need to contain the agricultural frontier is most urgent in the tropics, and while tropical deforestation continues to be commonly associated with vague notions of inefficient 'slash and burn' farming conducted by ignorant and overpopulated peasants, in most instances it is 'the lock that large landlords have on the productive resources' which is the decisive factor in continuing forest clearance (Thiesenhusen 1991: 8; Lohmann, 1993; Plant 1993). More industrial farming – whether more efficient or not – will only make this worse, not better.

This blind spot relates to some of the problems associated with environmental conservation in inequitable landscapes. The idea of conservation as an objective centred on natural areas, commonly through human-exclusive designations like national parks,[3] was exported throughout much of the world in the twentieth century. And while parks and protected areas do have a vital role in maintaining vulnerable species and ecosystems, they have often been established with little or no regard to how the exclusion or dislocation of local peoples can magnify existing inequalities (Kothari et al. 1995; Guha 1989); hence, when they are approached as a conservation objective in themselves, irrespective of disparities in land distribution, they often end up as 'islands' in the landscape and mired in a perpetually defensive state (the well-known 'fences and fines' predicament of forcibly keeping local people out). Thus, the argument that industrial farming will maximize non-agricultural land available to be protected can dangerously obscure how imbalances in land, control and resource consumption produce environmental degradation, and in practice narrowly conceived approaches to conservation amid uneven distributions of land not only raise social concerns but face a precarious future: 'islands' of nature are unlikely to be viable in the long term if they are situated in seas of injustice.

Another dubious assumption in the green case for industrial agriculture is how the meatification of human diets becomes normalized in taking the current demand for food as the basis for assessing the overall land-space needed for agriculture, when the current demand for food is not an immutable function of humanity's dietary needs.

Rather, as in all commodity markets, it is a function of how profit-seeking activity interacts with purchasing power, and just as there was nothing inevitable about the course of the global food economy there is nothing predestined about the seismic and continuing shift in global livestock populations and the per capita consumption of animal products. As was explained in Chapter 1, the increasing orientation of agricultural systems towards the production of meat and animal derivatives leads to an ever larger 'ecological hoofprint' given the inefficiencies of cycling grain through animals and the additional land-space devoted to producing animal feed, along with the enormous water and energy demands and waste generation problems of intensive livestock production. This development also poses deep environmental ethical questions about humanity's relationship with other species, particularly given the fast-spreading factory-farm model and the depths of physical and emotional suffering it entails, and about distributional equity. Thus, appeals to the yield efficiency of monoculture grain production as environmentally benign are ignoring the proverbial elephant – or literal cow, pig and chicken – in the room.

On this point, it bears noting that where the meatification of diets is not framed as an ethical non-issue, an assumed part of humans' entitlement as a species, it is often cloaked in the image of progress and 'higher-quality' diets (FAO 2002a; Gilland 2002). This image is often connected to the fact that the meat-intensive diets of the world's most affluent countries contain the highest per capita protein content in history, as well as to the biological point that protein from animal flesh is utilized by the human body more efficiently than is plant-based protein. And yet it is, as Rifkin (1992: 230) puts it, very much an 'artificial protein ladder' which humanity is ascending, as protein intake levels in meat-intensive diets vastly exceed our biological needs to the point of often having a detrimental effect on human health, making the issue of utilization efficiency superfluous. These negative health impacts are magnified when coupled with the high levels of saturated fats and cholesterol that are also characteristic of such diets, not to mention the concentrations of bio-accumulated pesticides, nitrites and pharmaceuticals and, in some countries, hormones. Because of this, as noted in Chapter 1, some major epidemiological studies have identified meat-centred diets as a significant contributing factor to a large range of chronic health problems, from cardiovascular disease (the leading cause of death in North America) to obesity (identified as a global health epidemic by the WHO), and urge instead that reducing or even eliminating animal products would offer immense public health benefits (e.g. Campbell and Campbell 2005; Chen et al. 1990). In short, the meatification of global diets is about neither biophysical necessity nor bettering the human condition: it is about supply and demand in an age of rising affluence.

When the green case for industrial agriculture is broken down, what emerges in its place is a very different understanding of efficiency and ecological rationality. Rather than expanding large high-input, high-yield monocultures, massive fossil fuel-powered farm machinery, factory farms and extraordinary productivity-to-farmer ratios, the aggregate ecological footprint associated with agriculture will tend to be reduced as the industrial grain-livestock complex is rolled back and societies move towards more equitable small-farm landscapes where labour-intensive multi-cropping prevails and production is more locally oriented. Such a transformation, especially where it can make inroads on the best-quality arable land, is where the potential to reduce the overall land-space needed for food production is truly harboured. This would in turn open the greatest possibilities for restoration, an objective that would have to balance the desirability of starting from the least suitable arable lands with the conservation biology goal of prioritizing landscape connectivity.

It is important to emphasize that the aim of progressively moving agricultural systems off the chemical and fossil energy treadmill and towards lower-input, labour-centred intensification and more biodiverse agriculture is not about going backwards to more 'primitive' approaches and rejecting modern science. On the contrary, to significantly increase the scale of organic and near-organic practices will require much more scientific research and training geared towards better understanding how agro-ecosystems operate and how key dynamics can be selectively enhanced. For instance, scientific research into the functional complementarities of various species can inform biological pest and disease control techniques, and scientific research into soil biochemistry and the retention and cycling of nutrients can commend specific cropping patterns and rotations, nitrogen-fixing plants and green manures to build up soil fertility. This agro-ecological science means pursuing applied research where farmers are key participants in the process of innovation, and where the traditional wisdoms accrued through generations of experimental, observational and working knowledge of a landscape are valued. This is not to uncritically venerate 'traditional' knowledge, which in many instances may have been ruptured by such things as the introduction of input packages and market pressures. It is instead to envision the building of more sustainable agricultural systems as 'a process of learning' rather than 'a simple model or package to be imposed'

(Pretty 1997: 249), where research is a two-way engagement between farmers and scientists, there are strong social (i.e. democratic and transparent) controls over research and the ensuing knowledge is entrusted as a public good.

To move towards a conception of agro-science where knowledge creation is driven by the interests of farmers and environmental and public health concerns and is held as a public good will involve resisting the corporate-driven science that is pushing hard in the opposite direction with deep-pocketed efforts: to simplify and industrialize farm systems; weave together dependence on interlocking inputs (agro-chemicals, seeds, fertilizers and livestock pharmaceuticals); privatize scientific knowledge in order to secure technological rents; and position farmers as recipients and consumers of these innovations. Because of the centrality of the commodified seed in locking in the input treadmill of industrial agriculture and supporting the monopolizing tendencies of agro-TNCs, the battle for the future of agricultural science will involve fighting for seeds as irrevocable common property. This means confronting the patentability of nature, the increasing agro-input TNC control over the world's seed stock, the scripting of national and multilateral intellectual property rights laws and the corporate penetration of public research agencies and universities.

Finally, in imagining scaling back the industrial grain-livestock complex there is the inescapable need to rethink the historically unprecedented scale of animal agriculture in human diets and economies, which is becoming ever more urgent as the world moves towards a global population of 9 billion people while, on average, sprinting towards more meat-centred consumption patterns. The mainstream environmental movement has historically given a great deal of attention to human population growth as a major environmental issue while for the most part ignoring the even faster growth of livestock populations, with the WorldWatch Institute a notable exception (Nierenberg 2005; WorldWatch 2004; Brown 1996; Durning and Brough 1991), having recently described the expanding 'human appetite for animal flesh' as 'a driving force behind virtually every major category of environmental damage now threatening the human future' (WorldWatch 2004: 12). The 'ecological hoofprint' of animal agriculture means that moving away from meat-centred consumption patterns is an elemental part of reducing humanity's collective space in the biosphere and leaving room for other species into this century, with well-balanced plant-centred diets also holding the additional promise of an array of public health benefits.

Obviously, such change is needed most in the rich, temperate countries at the centre of the global food economy where per capita consumption of animal products is the greatest, but this is also a major environment and development priority in the developing world (Goodland 1997), especially in the powerhouse economies of Asia, given the pace of dietary change and the magnitude of demand this entails.

Agriculture and social change In *The Earth*, the great nineteenth-century novelist Emile Zola described the French countryside in the throes of capitalist restructuring: 'in a word, the land had been turned into a bank, operated by financiers, exploited and cropped to the limit' (1980 [1887]: 454). Zola had, in other works, famously depicted the hellish industrial working conditions into which many peasants were being cast at the apogee of Europe's agrarian question, and while at times these miserable living and working conditions were shown to foment antisocial behaviours there was also a clear, underpinning hope about the revolutionary potential of this emergent working class, the key oppositional force capable of directing systemic change. In contrast, Zola's portrayal of the French peasantry in *The Earth* was both bleak and disparaging, hardly a force to be mobilized or a social class that could herald a more equitable and peaceful future. Such a picture was and remains a fairly customary one for socialist activists and intellectuals, shared with more celebratory accounts of capitalist restructuring (wherein the market is portrayed as the natural arbiter of productive resources), with peasant-based societies often seen as being at best relatively stagnant and at worst highly regressive and reactionary (Bernstein 2000). The associated belief is that the industrialization of agriculture is a necessary transformation in the course of human development, with the related image that this is 'releasing' people from the land and will eventually allow more beneficial forms of societal organization in the future, though there might initially be social costs both on the industrial farm, before farm labour is adequately mobilized, and off it, while those dislocated from farming adjust to the pressures of being thrust into new labour markets.

Despite the agrarian origins of capitalism, agriculture is the last major productive sector to have individual artisan producers fully 'proletarianized'; that is, dispossessed of the means of production and in waged servitude to capital as a property-less labour force. But agricultural systems around the world are clearly 'in the throes of the transition' to varying degrees, as 'the essence of proletarianization is in the loss of control over one's labour process and the alienation of the product of that labour' (Lewontin 2000: 94, 97). At the heart of the global food economy, the USA, industrial farms 'have among the highest per-unit levels of productivity, capital investment, and technological deployment' of any economic sector (Hahamovitch and Halpern 2004: 4). Many of the social costs associated with the proletarianization of agriculture and the declining viability of small farming have been discussed throughout this book, such as: the cost–price squeeze and spiralling indebtedness; farmer suicides from the USA to India; the exceptionally violent conditions of slaughterhouses and proliferating factory farms; the exploitation of non-unionized farm labour; the chronic exposure to toxic working conditions; and desperate waves of rural emigration into labour markets where un- and underemployment already widely exist on extraordinary scales. With respect to this last point, the biggest in human terms, *The Economist* (2006d: 6) has suggested that 'the global labour force has, in effect, doubled' as a result of the rapid restructuring, productivity increases and external market orientation of China, India and the former Soviet Union. And with a youthful world population still growing fast towards 9 billion this is poised to expand significantly again.

This growth, when set against the phenomenal concentrations of capital throughout the global economy, the hyper-productivity stemming from technological innovation and the global discipline on wages produced by the world's 'superfluous masses' means that dislocated small farmers in much of the world face even grimmer employment prospects than what Zola was describing in nineteenth-century Europe, exacerbated by the fact that developing countries lack much ability to export this problem. The spread of jobless de-peasantization together with the continuing 'commodification of everything' means that the challenge of meeting basic needs through the market with insecure and meagre incomes is likely to define the livelihood struggles in explosively growing urban and peri-urban areas of the developing world. As a result, it is not hard to see why some might be inclined to view small farming as a defensive social breakwater against rising waves of rural migration and the instability and suffering that adding another billion people to the 'planet of slums' by 2040 will entail (Davis 2006, 2004; UNHSP 2003).

If even progressive conceptions about the social significance of small farming stop here, however – and accept it as a historically outmoded residuum in the modern age – then there is not likely to be much momentum for agricultural policies that challenge the current trajectory, no matter how severe the immediate social costs. This amounts to a self-fulfilling prophecy, because in the absence of radical changes small farming is likely to become increasingly marginal to global food production and a decreasingly viable livelihood. Instead of transformative policies geared towards securing the future viability of small farming, as discussed in the second half of this chapter, the best we might then hope for are ameliorative supports for the displaced (e.g. public housing in cities, retraining pregrammes and income payments). Meanwhile, whatever pressure for change that might have come from the rising organic demand of health- and environment-conscious consumers could be satisfied by exploitative, labour-intensive corporate farms, with some finding solace in the expectation that this labour might become unionized if it is not already, or even one day socialized. The implication of all of this is that the battle for small farm-centred agrarian change can have popular and political appeal only if a hopeful social logic accompanies its ecological rationality. In other words, working to hold off a 'planet of slums' is not enough; to invigorate meaningful action small farmers must see themselves and have their livelihoods widely understood as an important part of building more socially just systems of production and sustaining healthy rural communities into the future.

A useful starting point towards such an understanding is 'the great question' of political economy – 'What have we to produce, and how?' (Kropotkin (1975 [1898]: 17). As Kropotkin noted a century ago, this question is relegated to the background in industrial capitalism and badly needs to be unpacked:

> Are the means now in use for satisfying human needs, under the present system of permanent division of functions and production for profits, really economical? Do they really lead to economy in the expenditure of human forces? Or, are they not mere wasteful survivals from a past that was plunged into darkness, ignorance and oppression, and never took into consideration the economical and social value of the human being? (ibid.: 193)

What we find, today as then, is that 'the expenditure of human forces' is governed by a logic that bears no relation to the objective of generating fulfilling, creative, socially meaningful and ecologically rational work. Instead, monopolizing tendencies concentrate control over what is produced and how into fewer and fewer hands; the overarching concerns of owners and managers are effective demand, profit growth and despatialized shareholders; the system is wired to insatiably expand apart from any environmental logic as demand is manufactured for so many unnecessary things and most environmental costs go unaccounted; and work is progressively atomized into smaller bits of the production process and separated from the natural world, with the industrial clock overriding daily and seasonal rhythms. In short, the problem of work today runs deeper than absolute labour surpluses and unequal compensation and crucially includes the systematic generation of alienating livelihoods. As Kropotkin (ibid.: 196) evocatively put it, 'What floods of useless sufferings deluge every so-called civilized land in the world!'[4]

In drawing attention to how the 'value of the human being' is lost in the ceaseless 'division of functions', Kropotkin made a persuasive case that the nature of work should be at the forefront of conceptions of human development. He argued that individual potential is maximized when people can combine 'brain work' and 'manual work', apply their varied talents and passions in a more fluid division of labour and have autonomy and creativity in their working lives, cultivated within the bounds of small-group activity and planning. In practice, this was seen to depend upon farms being closely connected to small factories and workshops in nearby towns and cities, with the assumption that societies would develop more rational ways of organizing production – from the development of what would later become known as 'appropriate technology' to the revaluation of time and leisure over money – when broadly educated working people drive the process of innovation and democratically govern diversified and localized production processes. Kropotkin (ibid.: 196) made it clear that this was looking forward rather than romanticizing the pre-industrial past: 'For thousands of years in succession, to grow one's food was the burden, almost the curse, of mankind. But it need be so no more.'

The old biases against farming as mind-numbing and back-breaking drudgery, which had long roots in human inequality and suffering, were an indictment of historic and existing social relations and magnified by soil-impoverishing techniques. An entirely different conception of farm work emerges, however, when production is governed by the sort of logic Kropotkin envisioned, something to be prized for its fusion of physical exertion in the open air with the varied intellectual challenges associated with managing agro-ecosystems and enhancing soils. Friedmann (2003), one of the great contemporary scholars of agrarian political economy, extends this case in the context of advancing agro-ecological knowledge and the potential benefits stemming from the democratic cross-fertilization of farming cultures, practices and ideas, helping us imagine how skilled, labour-intensive agriculture in biologically and culturally diverse communities could draw people closer to the biorhythms of the earth and even break down the rigid divide between work and leisure. Her buoyant image is of farming–gardening–playing coming together 'in the gardens of Gaia'.

Ultimately, the social case for small farming intersects strongly with the argument of the preceding section that biodiverse, labour-intensive and locally oriented small farms can have a major role in reducing humanity's ecological footprint. No 'expenditure of human forces' could be more important than producing healthy organic food from biodiverse, efficient and non-toxic agro-ecosystems, and no other sector can remotely compare with the extent to which small farming can generate dignified and fulfilling livelihoods. Compelling social and ecological reasons obviously do not give any assurance that necessary changes to the global food economy will have wide popular and political resonance, but as the battle for the future of farming plays out across multiple scales there is hope that they can help to give it strength.

The multiple scales of resistance and change

On 1 January 1994, the day NAFTA came into effect, the poorly equipped peasant forces of the Zapatista National Liberation Army seized the southern Mexican city of San Cristobal, five other towns and many large ranches in the state of Chiapas. NAFTA was both a literal and a symbolic target for the Zapatistas, as the legalized privatization of communal landholding, the institutionalization of market integration and the anticipated waves of subsidized US maize were seen as a direct assault on the 'people of maize' and another example of the way in which colonial injustices lived on in the exclusionary politics of the present. But after having won a national and even global stage, the Zapatistas quickly revealed themselves to be a new sort of revolutionary movement, rejecting both the use of force as a primary tool for social change and the quest for political power as a goal; as Subcomandante Marcos put it, 'the task of an armed movement should be to present the problem, and then step aside'.

In 1996, the Zapatistas hosted what they called the 'International Encounter Against Neoliberalism and For Humanity', framing the struggles between imposing and opposing neoliberal economic policies as a 'new world war that involves all of humanity'. The vision they articulated was of a plural but intertwined movement – 'one world where many worlds fit' – encapsulated in Marcos's elegant address to the meeting:

What do we do? Prepare a new schema to ease the anguish of not having a solution? Propose a global programme for world revolution? Or a Utopian theory to help us maintain a prudent distance from the reality that anguishes us? What we need is something smaller, but better. An echo that recognizes the existence of the other, and does not try to overpower or attempt to silence anyone. An echo of this rebel voice transforming itself and renewing itself in other voices. An echo of small magnitude, the local and the particular, reverberating in the echo of the great multitude, the intercontinental and galactic. An echo that turns itself into many voices, into a network of voices which, before the deafness of power, opts to speak for itself, recognizing itself in the diversity of the voices forming it. A network of voices that resists the war the powers wage upon us. A network of voices that not only speaks, but also struggles and resists for the sake of humanity and against neoliberalism.

The Zapatista uprising quickly resonated in Mexico and far beyond, and their creativity and ability to jump scales in analyses and actions, keeping both grounded locally while connected to broader struggles, was an important inspiration in the new waves of globally networked activism that emerged to challenge the triumphalism and institutional fortification of the Washington Consensus. When these waves coalesced in 1999 in the Battle of Seattle it was no coincidence that the collective movement borrowed a phrase from the revolutionary optimism of the Zapatistas: 'Another World is Possible!' Globally networked, multi-scaled activism has also picked up on the Zapatista idea of finding strength by approaching struggle as 'one no and many yeses'.

Confronting corporate-driven market integration An increasingly global lens is necessary to understand the problems facing farmers in most parts of the world. The idea of 'one no and many yeses' implies that while we need to oppose a universalizing system, alternatives will materialize in diverse ways as struggles to democratize and localize control play out within different environments, histories and cultural traditions. These objectives are entwined by the fact that the 'one no' must be confronted to expand the space for the 'many yeses' to flourish. For the global food economy the one inescapable 'no' is corporate-driven market integration and its institutional fortification, which entails 'greater freedom for the big to drive out the small, for forcing people everywhere to depend on distant global markets', and which could entail the 'triumph of inefficient and ecologically destructive monocultures over ecologically rational and sustainable farming practices' (Rosset 1999a: 17). Fortunately, as discussed in Chapter 4, the legitimacy of the WTO has been shaken repeatedly as the theorized promise underlying liberalization – that there are benign and efficient outcomes associated with rising trade – has become ever less tenable for the global food economy. It was also emphasized, however, that opponents of the WTO have struggled over the question of whether this is an irredeemable institution or a necessary and salvageable one, and hence whether the guiding objective should be to raze or to renovate it.

Would-be renovators argue that however uneven the WTO might be, developing countries nevertheless need a binding multilateral institution to discipline the economic policies of the rich countries of the world. In agriculture, much of this attention has been on the distortions posed by the large agro-subsidy regimes and selective protectionism in rich countries, given the obvious discord between the theory and practice of free trade that these represent, the 'hypocrisy' charge noted in Chapter 4. A related concern that was also discussed is that a lack of action would merely freeze the imbalances of the initial AoA and lead towards the proliferation of even worse bilateral trade negotiations. Conversely, the hope for the renovators is that developing countries could build on the assertiveness and alliances from Cancún onwards and profoundly reshape negotiating priorities and outcomes.

The broader prospect is that this might effectively renew the pressure for a 'New International Economic Order' that has been submerged for decades under the weight of debt, structural adjustment, liberalization and neoliberal orthodoxy. Since the problems of narrow export bases continue to define how the world's poorest countries are positioned within the global economy, and declining terms of trade for tropical commodities therefore constitute an enormous haemorrhaging of income (Khor 2002), a key motivation for renovating the WTO is that it could conceivably provide the necessary multilateral muscle to enact viable commodity agreements that coordinate supply management and lower and cap rich-country tariff peaks on processed goods. One notable case that some renovators have pointed to as evidence that the WTO can have a powerful role disciplining rich countries is the 2006 Appellate ruling against US cotton subsidies and in favour of the cotton exporters of West Africa, an issue that had received much international attention since the Cancún Ministerial (recalling how rich-country subsidies could not be challenged until the WTO's so-called 'peace clause' expired at the end of 2003).[5]

In agriculture, however, there is no easy way to reconcile the tensions between the negotiating priorities put forth by the governments of competitive agro-exporters like Brazil and Thailand with the food security and rural livelihood concerns upheld by governments such as those of India and the G-30+. The latter have argued for their need to maintain a significant measure of flexibility at the border into the future, advocating for the right of developing countries to self-identify a range of 'Special Products' and then have recourse to certain protective measures (i.e. special safeguards). But since the common justification for this argument – the multifunctional role that agriculture plays in societies – was first championed by the EU, it is highly implausible that this would be conceded in any substantive way for developing countries alone. For those opponents of the WTO who don't believe it is salvageable, these tensions underline a big reason why.

Critics like Rosset, Bello and La Via Campesina insist that the professed aim of renovating the WTO in the interests of the poor is misguided, and that its recurring conflicts and crises should be approached as a step towards razing either the AoA or the WTO in its entirety – getting the WTO *at least* out of agriculture' (Rosset 2006). One of the most basic ways in which the renovate perspective is seen to be mistaken is in falsely celebrating the developmental gains of rich-country subsidy discipline and improved market access for small farmers in developing countries, which are minimal, while downplaying the severe threats posed by rising imports and implicitly sanctifying the free-market approach to food security which, as this book has emphasized, is an especially specious theory in the face of the tropical commodities disaster, the persistent payments deficits of poor countries, the carbon/climate burden of large food miles and the potentially rising costs of long-distance food transport in an age of peak oil. So while the ruling against US cotton subsidies is likely to bring somewhat greater earnings for cotton exports from West Africa, at least in the short term, this could easily be read not as evidence of the potential value of the WTO but as justifying a harmful logic wherein some of the world's poorest and most food-insecure countries are encouraged to devote sizeable portions of their arable landscapes to producing a non-edible, low-value crop for distant markets amid rising food imports. And it should not be forgotten that record cotton exports were the flipside of the 1983/84 famine that struck many of these countries.

While the West African cotton case is somewhat unique in that small farmers are significantly plugged into the export network, in general the push to engage with the WTO as a means to reducing subsidies and selective protectionism in rich countries has very little to do with the interests of the large majority of farmers in the developing world, and is instead a reflection of the uneven political power exerted by large-scale farmers and commercial interests in the new agro-export powerhouses (which highlights the importance of understanding the divergent agrarian transformations discussed in Chapter 3 and who has actually benefited from successful agro-export growth). Rich-country subsidy regimes do of course play a very problematic role in shaping patterns of agricultural production and trade, but the call to renovate the WTO clearly risks making liberalization appear to be something that is primarily corrupted by governments. This can obscure the fact that the urgent need is not to *dismantle* public investment in agriculture but to *reshape* it, away from its present overwhelming and polarizing focus on industrial systems and towards research, extension, appropriate technology and other supports for low-input small farming, which will be especially crucial in a transitional period. Another serious criticism of the focus of the renovators is that due attention is not paid to the role that the WTO plays in securing 'the progressive removal of components of agricultural production from the control of the farmer via intervention in natural processes, starting with bio-engineered seeds, [and] complemented with a range of chemical and mechanical inputs' (McMichael 2004b: 9). Related to this is the extreme unlikelihood that the USA would allow the powers given to patent protections and biotechnology litigation to be weakened or over-ridden within the WTO.

In the end, the case for razing at least the AoA implores that we examine what would really be gained if emboldened alliances of developing countries could succeed in disciplining US and European subsidy regimes and tariff peaks, recognizing that this would almost surely come at the price of greater liberalization, and quite possibly the further erosion of democratic control over scientific innovation and patenting. The result would be a competitive playing field still weighted against the large majority of the world's small farmers, from the food import dependencies to the agro-export powerhouses, as they would remain widely constrained within inequitable landscapes and would not only have little prospect of state research and extension support but would face stronger coercions to produce in ways that are defined by agro-input TNCs bent on extracting greater technological rents from farming. Meanwhile, small farmers would be increasingly pitted against production from large-scale, industrialized farms processed, distributed and marketed by agro-food TNCs, which possess great abilities to influence dietary aspirations and whose power has accrued through time with the assistance of such things as the colonial structuring of agro-export landscapes, post-colonial food aid and decades of subsidized temperate exports.

The challenge to corporate-driven market integration has increasingly been framed in terms of 'food sovereignty'. The concept of food sovereignty does not romanticize the democratic character of nation-states, but emphasizes that decision-making and regulatory authority should be moving closer to farmers and farming systems, not farther from them. Berry (1995: 7) summarizes this basic logic with his usual eloquence:

> One thing at least should be obvious to us all: the whole human population of the world cannot live on imported food. Some people somewhere are going to have to grow the food. And wherever food is grown the growing of it will raise the same two questions: how do you preserve the land in use? And how do you preserve the people who use the land? The farther the food is transported, the harder it will be to answer those questions correctly. The correct answers will not come as the inevitable by-products of the aims, policies, and procedures of international trade, free or unfree. They cannot be legislated or imposed by international or national or state agencies. They can only be supplied locally, by skilled and highly motivated local farmers meeting as directly as possible the needs of informed local consumers.

Resistance to the WTO now has traction that runs much deeper than the formal negotiating impasses, and draws strength from a range of activism, research and advocacy organizations, critical scholarship, local and organic food networks, and most of all, farmers. Small farmers in developing countries are jumping scales of activism in new and powerful ways, with the best example being La Via Campesina, which has found common ground for alliance with family farm associations in the temperate world such as national farmers' unions, the US National Family Farm Coalition and the European Farmers Co-ordination (Rosset 2006; McMichael 2004b; Desmarais 2002). Where this resistance will lead remains to be seen, but if the AoA could be razed it would not only weaken the legal power of agro-TNCs over governments, opening greater space for alternatives, but would deal a heavy blow to the illusion that corporate-driven market integration is either inevitable or impregnable.

Land, seeds, body and mind One of the early catchphrases of sustainable development was 'think globally and act locally'. Similarly the notion of 'eating locally' has recently come into vogue for critics of the corporate-industrial food system and those working to build alternatives. This obviously presents very different challenges around the world, from temperate landscapes marked by industrial monocultures, immense productive surpluses, factory farms and overflowing, pseudo-diverse supermarkets, to the many tropical landscapes that are marked by narrow agro-export specializations, locally oriented small farming and rising net food import dependence. Amid this range of agricultural economies, the latter case is often held up by those who dismiss calls for a profound relocalization of global food production as being unrealistic (sometimes even describing it as iniquitous). Yet while a decline in food aid and cheap, subsidized food grain imports to poor countries would indeed tend to be most hurtful for the poorest within them in the short term, raising serious transitional issues, it is flawed circular logic to have one of the most grievous manifestations of the global food economy serve in defence of the free-market approach to food security.

What this points to most of all is that while the battle to contest the dominant actors and universalizing systemic logic transforming agriculture extends upwards in scale to the global level, to grasp the nature of agrarian struggle anywhere also crucially involves understanding how a given agrarian landscape has been structured, maintained and contested through time. Rosset (2006: 5) frames the land question in plain terms: 'Around the world, the poorest of the poor are the landless in rural areas, followed closely by the land-poor, those whose poor-quality plots are too small to support a family.' Ultimately, then, the work of building more localized, socially just and ecologically rational food economies hinges on challenging the grossly inequitable property rights that the global enclosures have wrought and working to reconfigure uneven landscapes – the ongoing struggle begun by the Diggers and Levellers nearly four centuries ago. This will entail the concerted efforts of farmers, their associations, social and environmental movements and concerned scholars to 'put land reform back on the national and international agendas of development' (Leonard and Manahan 2004: 14), as a report by the Land Research Action Network insists.[6]

Land reform tends to be discussed almost entirely in the context of the developing world, with most action on land questions in developed countries focused in a defensive way, bent on preserving family farms, defending remaining public lands against encroachment and, in some places, stopping urban sprawl. But, as was discussed in Chapter 2, misshapen subsidy regimes, tax systems and economies of scale have fuelled ongoing processes of capitalization, indebtedness, bankruptcies, displacement and land accumulation in much of the temperate world, making farming nearly impossible to enter into and difficult to stay in, even with a family history on the land. For the health of its own rural communities, agrarian landscapes and water resources, as well as for the interests of farmers around the world, a nation like 'the US needs land reform as much as does Latin America' (George 1990: 197), though pushing the matter of land redistribution into the realms of public consciousness and policy debates in temperate grain-livestock landscapes is an even steeper uphill battle than it is in developing countries.

While calls for land reform in the developing world have long been linked to struggles for social justice, conceptually and in practice, Marxist critics have warned that attention must be paid to differentiated rural classes. The fear is that populist calls for land reform can easily lead to a slippery slope of land accumulation from below (i.e. by wealthier small or medium-sized farmers), which is especially likely in the case of the market-led land reforms promoted by the World Bank (Borras 2003). This is an important caution, but alternatively there is no single model for how radical land reform might look. My own ideological preference is for historically and culturally contingent forms of community or state-held property with secure, renewable leases for farmers. This system of ownership plus long-term leases would be aimed at: providing a check against dispossession and land accumulation over time; giving farmers motivation and legal protections to cultivate with the long-term health of the land in mind; and placing authority with local, democratic bodies to revoke tenure when practices threaten the ecological health of the landscape. But however radical land reforms are conceived, pressure to actualize them will generally depend on the strength of social movements on the ground, as is now emanating from the mobilizations of the landless peoples, small farmers and indigenous groups across much of Latin America, the largest and most famous of which is the MST in Brazil. This pressure is helping to change the political dynamics across the continent, and with this might soon begin to significantly reshape the agrarian landscapes as well (Veltmeyer 2005; Branford and Rocha 2002).

Inspiration for the interwoven possibilities of land reform, the localization of food production and the shift from industrial to organic agriculture can also be found in Latin America and the Caribbean in the now widely cited case of Cuba, though Cuba's unique and urgent circumstances led to unusually strong pressure from above for such a transformation. Until very recently, Cuba's agricultural economy was a classic tropical monoculture. Its agrarian landscape had been forged by King Sugar and slavery, and the vast inequalities of the plantation system produced fertile ground for Castro's guerrilla army in rural areas in the late 1950s. Revolutionary Cuba challenged its historical inheritance in many ways, including the expropriation and conversion of all largeholdings into state farms, but its incorporation within the Soviet sphere and favourable terms of trade kept the agricultural sector bound in a neocolonial relationship – sugar production continued to consume roughly three times as much land as did domestic food crops, food imports grew and the industrialization of the model increased dependence on imported agro-chemicals, machinery and petroleum (with the significant exception being that, as noted in Chapter 3, inflated sugar earnings were not dominated by a small landed elite or foreign companies but were used to subsidize Cuba's social achievements in areas like health and education) (Funes et al. 2002; Rosset 2000).

The failings of Cuba's agro-export landscape were quickly exposed with the demise of the Soviet empire, in particular the fragility of food security, the alienation of workers on collective farms and the heavy dependence on imported inputs. The evaporation of favourable terms of trade for sugar combined with the crippling US embargo to produce dire food and input shortages in the 'Special Period', and almost overnight the importation of basic foodstuffs and oil fell by more than half and that of fertilizers and pesticides by 80 per cent. In 1993–94, when the economic crisis was most intense, the average Cuban citizen lost between 10 and 20 pounds, with the daily caloric intake per capita reaching a low ebb of little more than two-thirds of the WHO's recommended levels. In the face of this perilous food insecurity, the government quickly recognized that the large, state-run industrialized farms had become obsolete amid shortages of foreign exchange, spare parts, agro-inputs and petroleum, and that small farms were much more adaptable to the lower-input techniques needed in this new conjuncture. In response, an extensive divestment of state lands was begun in 1993, with laws to limit the size of landholdings, and market mechanisms (e.g. new farmers' markets) were mixed with state controls. The move to lower-input techniques was supported by shifting some of the state's considerable scientific capacities into agro-ecological research and extension, accentuating the important role for government supports in this transition, and the substitution of draught animals for fossil fuels accompanied the economically motivated shift to organic techniques. The net result has been a remarkable improvement in national food production and food security, and the speed and success of this transformation help to challenge the dominant narratives of what is possible (Funes et al. 2002; Rosset 2000).

Though Cuba's organic transformation was born of a particular exigency, a key stress that precipitated it was the sudden spike in the cost of its agro-inputs and petroleum imports, and this might soon start impacting much more widely. The intersection of the age of peak oil and high fuel costs with large and rising food miles is one of the great contradictions of the global food economy, given how fossil fuels are embedded in extensive transport, processing, refrigeration, farm machinery and agro-chemicals and fertilizers, and even if we assume that carbon emissions and toxic burdens will continue to be largely externalized in cost-accounting systems this is a contradiction that could soon begin to fracture the distorted competitive advantage of industrial foods (which the prospect of biofuels is not likely to dint).[7]

Another major contradiction in industrial agriculture relates to how agro-input TNCs, amid the breakneck corporate clustering and interlocking of seeds with agro-chemicals, have sought to circumvent public debates and institutions in propelling GMOs into fields and food supplies. Because GMOs pose uncertain but potentially serious hazards to food safety and ecosystem health, and assail widespread conceptions of seeds as common property, their propulsion has involved a range of devious tactics such as 'accidental releases' and intensive lobbying efforts to oppose food labelling, independent testing and public disclosure standards. In the few agricultural systems where this political influence has been successful, most pivotally the USA, the result is that 'consumers are eating GMOs, whether they know it or not', lacking even the right to know, and hundreds of open-field test plots grow transgenic crops containing 'drugs, human genes, animal vaccines, and industrial chemicals, without sufficient safeguards to protect nearby food crops' (Cummings 2005: 30). This multifaceted threat to farmers' control, agrarian landscapes, consumers' rights (including over their own bodies) and democratic processes has produced growing opposition, to such an extent that 'the fight against GM foods' has been described as 'the leading edge of a struggle by farmers and the general public against complete corporate control of the food system' (Magdoff et al. 2000: 17).

This struggle has played out in different ways. Effective popular education and political activism throughout Europe have worked to make food labelling mandatory and contest the importation of GM and non-labelled foods, with an important by-product being the tension this issue has created between the USA and the EU at WTO negotiations. Educational and political campaigning have also been complemented by direct actions, with the burning and uprooting of GM crops in nations as diverse as England, India and Brazil serving not only to contain the genetic contamination of landscapes but to draw much attention to the issue – Lynas (2004) provides an excellent summary of this in the English case. While agro-input TNCs such as Monsanto have sought to portray these as acts of 'eco-terrorism', such claims are not merely made on a feeble footing; they are, for Monsanto, like treading on quicksand: after all, not many reasonable people would describe as terrorism efforts to defend farmers' age-old rights to save seeds, ensure that there are strong social controls over scientific research and oppose the patentability of life and the potentially dangerous (but profitable) mutations of genes, while trusting in the benevolence of secretive corporate research, corporate voluntarism and patent lawyers to lead to the greater social and ecological good.

From seasonal cultivation routines to harvest to preparation to mealtime, food has long been a central part of cultural identities, and a major aspect of the escalating power of agro-TNCs lies in their extraordinary ability to sever both the material and conceptual links between farmers and consumers and replace these with opaque webs of sourcing, processing, distributing, retailing and branding while, as discussed at the outset of this chapter, managing to naturalize this. Conversely, a big reason for the potency of GMOs galvanizing resistance to corporate-industrial agriculture lies in how these issues connect the concerns of farmers and consumers while, for many, shattering illusions of the trajectory being directed by benign or inevitable forces. This, in turn, highlights the need for education and activism to (re)position food in people's everyday consciousness, as well as the unique power that food can have in helping people to appreciate systemic contradictions. Because however much food is transformed into a despatialized commodity it will always remain a biophysical encounter connecting the human body to systems of production, nature and other species in a very intimate way, and when people come to grasp the system in their food this awareness reverberates multiple times every day.

Though GMOs cast this into clear relief, there is an even more compelling way in which the system of corporate-industrial agriculture can be felt in this biophysical encounter of food. As has been emphasized throughout this book, the soaring scale of farm animal production and meatification of diets is central to understanding the global food economy. Rising awareness about some of the problems with the industrial grain-livestock complex is beginning to be seen in such things as popular reactions to mad cow and avian flu scares around the world, grassroots struggles against the spread of factory farming into communities and increasing attention to the brutal working conditions on the lines of modern slaughterhouses and meat-packing plants. One striking portrayal of the problems with industrial livestock production is the computer animation *The Meatrix* (www. meatrix.com), devised by a US consumer education organization. But as yet much of this concern might be characterized as being focused on technological Band-Aids for food safety and epidemiological threats, narrow NIMBY (not in my backyard) resentment about location and protections for vulnerable workers, rather than pointing towards more transformative ethical and structural change. For instance, in his best-selling account of the fast-food industry, Schlosser (2002) gives a generally searing indictment of industrial agriculture in the USA, yet ends up emphasizing the replacement of upwardly mobile union packing jobs with the exploitation of non-unionized and migrant workers, going so far as to romanticize the lost 'soul' of the 'Jungle' in Chicago not long after describing how the beginning of those early slaughterhouse lines saw workers smashing cows over the head with a sledgehammer.

There is an urgent need to foster further popular understanding of the problems with the industrial grain-livestock complex, the ethical and environmental implications of its ecological hoofprint and the exploding scale of farm animal suffering. Walker (1988: 9) conveys this second point powerfully at the end of a short essay:

> ... we are used to drinking milk from containers showing 'con-
> tented' cows, whose real lives we want to hear nothing about, eating
> eggs and drumsticks from 'happy' hens, and munching hamburgers
> advertised by bulls of integrity who seem to command their fate.
>
> As we talked of freedom and justice one day for all, we sat down
> to steaks. I am eating misery, I thought, as I took the first bite. And
> spit it out.

Grasping this miserable flesh as the embodiment of the systemic logic of corporate-industrial agriculture, while simultaneously widening our sphere of moral concern as a species, leads towards the sort of revolutionary consciousness that could inspire profound changes, individually and collectively.

Notes

1 The *Economist* (2006d: 75) tries to depict food miles as being a 'misleading' concept using some rather obfuscating feats of logic, including the obvious suggestion that 'a mile travelled by a large truck full of groceries is not the same as a mile travelled by an SUV carrying a bag of salad', as well as effectively comparing apples shipped in bulk containers across large distances to oranges produced locally and then sold and transported one by one. Of course, much local food still gets from field to plate in ecologically irrational ways, but surely these straw-men arguments point to other problems in local transportation planning and market incentives rather than detracting from the basic logic that the localization of food economies will tend to reduce food miles and in turn reduce the fossil fuel consumption and carbon emissions associated with agriculture.

2 The summary discussion of agroecology and small-farm efficiency draws mainly from Gleissman (2000, 1997), Altieri (1998, 1995), Rosset (1999a, b), Pretty (1997, 1995), Netting (1993), Paoletti et al. (1992), and Altieri and Hecht (1990).

3 The contemporary model of national parks and protected areas arose in late-nineteenth-century American conservation struggles which fought to save some wild or pristine lands from ranching, mining, logging and dams as industrial society pushed rapidly westward, and the struggle for human-exclusive spaces became a central priority of mainstream environmentalism in the USA. But this view of nature and approach to conservation left out the First Nations inhabiting the land, who had no cultural conception of wilderness. It also left out the violence of conquest, and in a way this historical and cultural myopia helped to fortify the image of empty lands that was part of the colonizing narrative.

4 This is not to downplay the fact that some powerful ideological constructs work to obscure this, such as class stratification as 'meritocracy', wages as a reasonable measure of the worth of labour and freedom as the right to consume whatever you can afford. Further, consumption in countless material and cultural forms (e.g. television programmes, movies, sport, celebrity fixations, video games, etc.) provides a degree of inoculation against stultifying work, and clearly some people find fulfilment and stimulation in narrowly defined functions irrespective of whether or not the production process in which they are embedded serves a social good or whether the environmental implications are understood. Nevertheless, we can assume that widespread disaffection with the nature of work is one important seam for rupturing the hegemony of industrial capitalism.

5 The USA and the EU had lobbied to extend the peace clause and even went so far as to suggest that WTO-sanctioned challenges to their subsidy regimes would sour future negotiations.

6 The Land Research Action Network links together activists and researchers to learn from and disseminate lessons about struggles occurring in different places.

7 While biofuels have the great advantages of being a renewable source

of energy and burning much more cleanly than fossil fuels, the prospect of deriving biofuels from things like corn and sugar monocultures on a scale that could significantly affect the world's energy supply without radically reducing demand is a fanciful one. And if attempts are made to significantly expand biofuel production to substitute for fossil fuels on any sizeable scale without reducing energy demand and the volume of feed grains cycled through livestock it is hard to imagine any space being left in the biosphere for self-organizing ecosystems and other species.

Bibliography

Agarwal, B. (1994) *A Field of One's Own: Gender and Land Rights in South Asia*, Cambridge: Cambridge University Press.

Altieri, M. A. (1995) *Agroecology: The Science of Sustainable Agriculture*, Boulder, CO: Westview.

— (1998) 'Ecological impacts of industrial agriculture and the possibilities for truly sustainable farming', *Monthly Review*, 50(3).

Altieri, M. and S. H. Hecht (eds) (1990) *Agroecology and Small Farm Development*, Florida: CRC.

Amin, S. (2003) 'World poverty: pauperization and capital accumulation', *Monthly Review*, 55(5).

Andrea, G. and B. Beckman (1985) *The Wheat Trap: Bread and Underdevelopment in Nigeria*, London: Zed Books.

Araghi, F. (2000) 'The great global enclosure of our times: peasants and the agrarian question at the end of the twentieth century', in F. Magdoff, J. B. Foster and F. H. Buttel (eds), *Hungry for Profit: The Agribusiness Threat to Farmers, Food, and the Environment*, New York: Monthly Review Press, pp. 145–60.

Athanasiou, T. and P. Baer (2002) *Dead Heat: Global Justice and Global Warming*, New York: Seven Stories Press.

Avery, A. and D. Avery (2003) 'High-yield conservation: more food and environmental quality through intensive agriculture', in R. E. Meiners and B. Yandle (eds), *Agricultural Policy and the Environment*, Lanham, MD: Rowman & Littlefield, pp. 135–50.

Avery, D. (1999) 'Intensive farming and biotechnology: saving people and wildlife in the 21st century', in G. Tansey and J. D'Silva (eds), *The Meat Business: Devouring a Hungry Planet*, New York: St Martin's Press, pp. 15–22.

Badiane, O. and M. Kherallah (1999) 'Market liberalisation and the poor', *Quarterly Journal of International Agriculture*, 38(4).

Bailey, R. (ed.) (2002) *Global Warming and Other Eco-Myths: How the Environmental Movement Uses False Science to Scare Us to Death*, Roseville, CA: Prima Publishing.

Bangasser, P. E. (2000) *The ILO and the Informal Sector: An Institutional History*, Geneva: ILO.

Barboza, D. (2001) 'Meatpackers' profits hinge on pool of immigrant labor', *New York Times*, 21 December.

Barkin, D. (1987) 'The end of food self-sufficiency in Mexico', *Latin American Perspectives*, 14(3).

Barlow, M. and T. Clarke (2002) *Blue Gold: The Battle against Corporate Theft of the World's Water*, Toronto: McLelland and Stewart.

Barnard, N. D. (1993) *Food for Life: How the New Four Food Groups Can Save Your Life*, New York: Harmony.

Barndt, D. (2002) *Tangled Routes: Women and Globalization on the Tomato Trail*, Toronto: Garamond.

Barraclough, S. L. (1994) 'The legacy of Latin American land reform', *NACLA Report on the Americas*, 28(6).

Barry, T. (1987) *Roots of Rebellion: Land and Hunger in Central America*, Boston, MA: South End Press.

— (1995) *Zapata's Revenge: Free Trade and the Farm Crisis in Mexico*, Boston, MA: South End Press.

Becker, E. (2001) 'Far from dead, subsidies fuel big farms', *New York Times*, 14 May.

— (2002) 'US defends its farm subsidies against rising foreign criticism', *New York Times*, 27 June.

Bello, W. (1994) *Dark Victory: The United States, Structural Adjustment and Global Poverty*, San Francisco, CA: Food First.

— (2002a) *De-globalization: Ideas for Running the World Economy*, London: Zed Books.

— (2002b) 'Lack of transparency in the WTO', *Development Dialogue*, 1.

— (2003a) 'There is life after Cancún', *Focus on Trade*, 23 September, <http://focusweb.org/content/index2.php?option=com_content&do_pdf=1&id=159>.

— (2003b) 'The meaning of Cancún', *Focus on Trade*, 93, September, <www.focusweb.org/publications/FOT%20pdf/fot93.pdf>.

— (2003c) 'Why a derailed WTO Ministerial is the best outcome for the South', *Focus on Trade*, 91, September, <www.focusweb.org/publications/FOT%20pdf/fot91.pdf>.

Bente, L. and S. A. Gerrior (2002) 'Selected food highlights of the 20th century: US food supply series', *Family Economics and Nutrition Review*, 14(1).

Berlan, J.-P. (1991) 'The historical roots of the present agricultural crisis', in W. H. Friedland, L. Busch, F. H. Buttel and A. P. Rudy (eds), *Towards a New Political Economy of Agriculture*, Boulder, CO: Westview, pp. 115–36.

Bernstein, H. (1990) 'Taking the part of peasants', in H. Bernstein, B. Crow, M. Mackintosh and C. Martin (eds), *The Food Question: Profits versus People*, New York: Monthly Review Press, pp. 69–79.

— (2000) '"The peasantry" in global capitalism: who, where and why?', in L. Panitch and C. Leys (eds), *Socialist Register 2001: Working Classes, Global Realities*, New York: Monthly Review Press, pp. 25–51.

Berry, R. A. (1997) 'Agrarian reform, land distribution, and small-farm policy as preventive of humanitarian emergencies', Paper presented to the Political Economy of Humanitarian Emergencies conference, Oxford, July.

Berry, W. (1995) *Another Turn of the Crank*, Washington, DC: Counterpoint.

Binswager, H. P. and K. Deininger (1997) 'Explaining agricultural and agrarian policies in developing countries', *Journal of Economic Literature*, 35(4).

Borras, S. M. (2003) 'Questioning market-led agrarian reform: experiences from Brazil, Colombia and South Africa', *Journal of Agrarian Change*, 3(3).

Branford, S. and J. Rocha (2002) *Cutting the Wire: The Story of the Landless Movement in Brazil*, New York: Latin American Bureau.

Brown, D. (1970) *Bury My Heart at Wounded Knee: An Indian History of the American West*, New York: Holt, Rinehart & Winston.

Brown, L. R. (1996) *Tough Choices: Facing the Challenge of Food Scarcity*, New York: W. W. Norton.

Brown, L. R. and H. Kane (1994) *Full House: Reassessing the Earth's Population Carrying Capacity*, New York: W. W. Norton.

Bumpers, D. (1998) Statement given to the Hearings based on the Prepared Statement of the Secretary of Agriculture to the US Senate, Washington, DC: US Government Printing Office, February.

Burniaux, J. M., F. Delorme, I. Lienert, J. P. Martin and P. Hoeller (1988) 'Quantifying the economy-wide effects of agricultural policies: a general equilibrium approach', OECD Working Paper no. 55, Paris.

Buttel, F. H. (2003) 'Internalizing the society costs of agricultural production', *Plant Physiology*, 133(12).

Buttel, F. H. and A. Hirata (2003) 'The "gene revolution" in global perspective: a reconsideration of the global adoption and diffusion of GM crop varieties, 1996–2002', Paper no. 9, Program on Agricultural Technology Studies Staff Paper Series, University of Wisconsin-Madison.

Campbell, T. C. with T. M. Campbell (2005) *The China Study: The Most Comprehensive Study of Nutrition Ever Conducted and the Startling Implications for Diet, Weight Loss and Long-term Health*, Dallas, TX: BenBella Books.

Carson, R. (1962) *Silent Spring*, New York: Houghton Mifflin.

Carter, M. R. and B. L. Barham (1996) 'Level playing field and *laissez faire*: post-liberal development strategy in inegalitarian agrarian economies', *World Development*, 24(7).

Carter, M. R. and J. Coles (1998) 'Inequality-reducing growth in agriculture: a market-friendly policy agenda', in N. Birdsall, C. Graham and R. H. Sabot (eds), *Beyond Tradeoffs: Market Reform and Equitable Growth in Latin America*, Washington, DC: Inter-American Development Bank and the Brookings Institute Press, pp. 147–82.

Carter, M. R., B. L. Barham and D. Mesbah (1996) 'Agricultural export booms and the rural poor in Chile, Guatemala and Paraguay', *Latin American Research Review*, 31(1).

Castells, M. (1993) 'The informational economy and the new international division of labour', in M. Carnoy, M. Castells, S. S. Cohen and F. H. Cardoso (eds), *The New Global Economy in the Information Age: Reflections on Our Changing World*, University Park: Pennsylvania State University Press, pp. 15–45.

Chaliand, G. (1977) *Revolution in the Third World*, New York: Viking.

Chang, H.-J. (2003) 'Kicking away the ladder: neoliberals rewrite history', *Monthly Review*, 54(8).

Chen, J., T. C. Campbell, J. Li and R. Peto (1990) *Diet, Lifestyle, and Mortality in China: A Study of the Characteristics of 65 Countries*, New York: Oxford University Press.

Chossudovsky, M. (1998) *The Globalization of Poverty: Impacts of IMF and World Bank Reforms*, London: Zed Books.

Chrispeels, M. J. (2000) 'Biotechnology and the poor', *Plant Physiology*, 124(9).

Clapp, J. (2006) 'WTO agriculture negotiations: implications for the Global South', *Third World Quarterly*, 27(4).

Cochrane, W. W. (2003) *The Curse of American Agricultural Abundance: A Sustainable Solution*, Lincoln: University of Nebraska Press.

Cockburn, A. (1996) 'A short meat-oriented history of the world', *New Left Review*, 215.

Conway, G. (1997) *The Doubly Green Revolution: Food for All in the 21st Century*, London: Penguin.

Cook, I. (1994) 'New fruits and vanity: symbolic production in the global food economy', in A. Bonanno, L. Busch, W. H. Friedland, L. Gouveia and E. Mingione (eds), *From Columbus to ConAgra: The Globalization of Agriculture and Food*, Lawrence: University of Kansas Press, pp. 232–48.

Cornia, G. A. (1985) 'Farm size, land yields and the agricultural production function: an analysis for fifteen developing countries', *World Development*, 13(4).

Crister, G. (2004) *Fat Land: How Americans Became the Fattest People in the World*, New York: Houghton Mifflin.

Cronon, W. (1991) *Nature's Metropolis: Chicago and the Great West*, New York: W. W. Norton.

Crosby, A. W. (1972) *The Columbian Exchange: Biological and Cultural Consequences of 1492*, Westport, CT: Greenwood.

— (1986) *Ecological Imperialism: The Biological Expansion of Europe, 900–1900*, Cambridge: Cambridge University Press.

Cummings, C. H. (2005) 'Trespass', *WorldWatch Magazine*, January/February.

Das, B. L. (1998) *Trade and Development Issues and the World Trade Organization*, vol. 1: *An Introduction to the WTO Agreements*, vol. 2: *The WTO Agreements: Deficiencies, Imbalances, and Required Changes*, London: Zed Books.

— (1999) *The World Trade Organization: A Guide to the New Framework for International Trade*, London: Zed Books.

— (2003a) 'Salvaging WTO from Cancun collapse', *Third World Network*, 20 September, <www.twnside.org.sg/title/twninfo78.htm>.

— (2003b) *The WTO and the Multilateral Trading System: Past, Present, and Future*, London: Zed Books.

Davis, K. (1995) 'Thinking like a chicken: farm animals and the feminine connection', in C. J. Adams and J. Donovan (eds), *Animals and Women:*

Feminist Theoretical Perspectives, Durham, NC: Duke University Press, pp. 192–212.

— (1996) *Prisoned Chickens, Poisoned Eggs: An Inside Look at the Modern Poultry Industry*, Summertown, TE: Book Publishing Company.

— (2005) 'The avian flu crisis', Address given at the Thinking about Animals in Canada conference, St Catherine's: Brock University, 24/25 February.

Davis, M. (2004) 'Planet of slums: urban involution and the informal proletariat', *New Left Review*, 26, March/April.

— (2005) *The Monster at Our Door: The Global Threat of Avian Flu*, New York: New Press.

— (2006) *Planet of Slums*, London: Verso.

De Haan, C., S. Henning and H. Blackburn (1997) *Livestock and the Environment: Finding a Balance*, Brussels: Commission of the European Communities, Food and Agriculture Organization, United States Agency for International Development, World Bank.

Deininger, K. and H. Binswager (2002) 'The evolution of the World Bank's land policy', in A. de Janvry, G. Gordillo, J. P. Platteau and E. Sadoulet (eds), *Access to Land, Rural Poverty, and Public Action*, New York: Oxford University Press.

Delgado, C. L. and C. A. Narrod (2002) 'Impact of changing market forces and policies on structural change in the livestock industries of selected Fast-growing Developing Countries', Report to the FAO by IFPRI, <www.fao.org/wairdocs/LEAD/X6115E/x6115e00.HTM>.

Delgado, C. L, M. Rosegrant, H. Steinfeld, S. Ehui and C. Courbois (1999) 'Livestock to 2020 – the next food revolution', Food, Agriculture, and Environment Discussion Paper 28, Washington, DC: International Food Policy Research Institute.

Desmarais, A.-A. (2002) 'The Vía Campesina: consolidating an international peasant and farm movement', *Journal of Peasant Studies*, 29(2).

Duncan, C. (1996) *The Centrality of Agriculture: Between Humankind and the Rest of Nature*, Montreal and Kingston: McGill-Queen's University Press.

Dundon, S. J. (2003) 'Agricultural ethics and multifunctionality are unavoidable', *Plant Physiology*, 133(12).

Durning, A. B. and H. B. Brough (1991) *Taking Stock: Animal Farming and the Environment*, WorldWatch Paper no. 103, Washington, DC.

EACCNCE (Economic Advisory Council Committee on Nutrition in the Colonial Empire) (1939) 'Nutrition in the Colonial Empire: first report', Report Presented to Parliament by Command of His Majesty, London.

Economist (2003) 'Spoilt for choice: a survey of food', *The Economist* (special section), 13 December.

— (2005) 'Europe's farm follies: why the EU retains its strange fondness for farm subsidies', *The Economist*, 8 December.

— (2006a) 'The future of globalisation', *The Economist*, 29 July.

— (2006b) 'In the twilight of Doha', *The Economist*, 29 July.

— (2006c) 'The new titans', *The Economist*, 16 September.

— (2006d) 'Food politics: voting with your trolley', *The Economist*, 9 December.

Ellis, R. (2003) *The Empty Ocean: Plundering the World's Marine Life*, Washington, DC: Island Press.

ETC Group (Action Group on Erosion, Technology, and Concentration) (2005a) 'Who owns whom? seed industry concentration – 2005', *Communiqué*, 90, September/October.

— (2005b) 'Oligopoly, Inc. 2005 concentration in corporate power', *Communiqué*, 91, November/December.

Faber, D. (1993) *Environment under Fire: Imperialism and the Ecological Crisis in Central America*, New York: Monthly Review Press..

Fagan, R. (1997) 'Local food/global food: globalization and local restructuring', in R. Lee and J. Wills (eds), *Geographies of Economies*, London: Edward Arnold, pp. 197–208.

Fanon, F. (1982 [1963]) *The Wretched of the Earth*, New York: Grove Press.

FAO (Food and Agricultural Organization of the United Nations) Statistics database (FAOSTAT), <http://faostat.fao.org/default.aspx>.

— (1997) *The State of the World's Plant Genetic Resources for Food and Agriculture*, Rome: FAO.

— (1998) 'FAO: conventional tilling severely erodes the soil; new concepts for soil conservation required', FAO press release 98/42, Rome: FAO.

— (2002a) *World Agriculture: Towards 2015/2030 – Summary Report*, Rome: FAO.

— (2002b) 'Meat and meat products', *FAO Food Outlook*, 4, Rome: FAO.

— (2002c) *The State of World Fisheries and Aquaculture*, Rome: FAO.

— (2003) *The State of Food Insecurity in the World 2003: Monitoring progress towards the World Food Summit and Millennium Development Goals*, Rome: FAO.

— (2006) 'Better water management means a healthier environment', FAO newsroom, Rome, <www.fao.org/newsroom/en/focus/2006/1000252/article_1000253en.html>.

Fernández-Armesto, F. (2002) *Near a Thousand Tables: A History of Food*, New York: Free Press.

Foster, J. B. (1999) *The Vulnerable Planet: A Short Economic History of the Environment*, 2nd edn, New York: Monthly Review Press.

Foster, J. B. and F. Magdoff (2000) 'Liebig, Marx, and the depletion of soil fertility: relevance for today's agriculture', in F. Magdoff, J. B. Foster and F. H. Buttel (eds), *Hungry for Profit: The Agribusiness Threat to Farmers, Food, and the Environment*, New York: Monthly Review Press, pp. 43–60.

Fox, M. (1999) 'American agriculture's ethical crossroads', in G. Tansey and J. D'Silva (eds), *The Meat Business: Devouring a Hungry Planet*, New York: St Martin's Press, pp. 25–42.

Friedmann, H. (1990) 'The origins of Third World food dependence', in H. Bernstein, B. Crow, M. Mackintosh and C. Martin (eds), *The Food Question: Profits versus People*, New York: Monthly Review Press, pp. 13–31.

— (1993) 'The political economy of food: a global crisis', *New Left Review*, 197.

— (1994) 'Shaky foundations of the world food economy', in P. McMichael (ed.), *The Global Restructuring of Agro-food Systems*, Ithaca, NY: Cornell University Press, pp. 258–76.

— (2000) 'What on earth is the modern world-system? Foodgetting and territory in the modern era and beyond', *Journal of World Systems Research*, 6(2).

— (2003) 'Eating in the gardens of Gaia: envisioning polycultural communities', in J. Adams (ed.), *Fighting for the Farm: Rural America Transformed*, Philadelphia: University of Pennsylvania Press, pp. 252–73.

— (2004) 'Feeding the empire: the pathologies of globalized agriculture', in L. Panitch and C. Leys (eds), *The Empire Reloaded: Socialist Register*, New York: Monthly Review Press, pp. 124–43.

Friedmann, H. and P. McMichael (1989) 'Agriculture and the state system', *Sociologia Ruralis*, 29(2).

Frontline (2006) 'The price of reforms', *Frontline*, 23(12).

Funes, F., L. García, M. Bourque, N. Pérez and P. Rosset (eds) (2002) *Sustainable Agriculture and Resistance: Transforming Food Production in Cuba*, Oakland, CA: Food First Books.

Galeano, E. (1973) *Open Veins of Latin America: Five Centuries of the Pillage of a Continent*, New York: Monthly Review Press..

Gandhi, M. (1999) 'Factory farming and the meat industry in India', in G. Tansey and J. D'Silva (eds), *The Meat Business: Devouring a Hungry Planet*, New York: St Martin's Press, pp. 92–100.

Garst, R. and T. Barry (1990) *Feeding the Crisis: US Food Aid and Farm Policy in Central America*, Lincoln: University of Nebraska Press.

George, S. (1990) *Ill Fares the Land: Essays on Food, Hunger, and Power*, London: Penguin.

Gilland, B. (2002) 'World population and food supply: can food production keep pace with population growth in the next half-century?', *Food Policy*, 27(1).

Gleissman, S. R. (1997) *Agroecology: Ecological Processes in Agriculture*, Ann Arbor: University of Michigan Press.

— (ed.) (2000) *Agroecosystem Sustainability: Developing Practical Strategies*, Boca Raton, FL: CRC Press.

Glickman, D. (1998a) Statement of the Secretary of Agriculture to the US Senate, Washington, DC: US Government Printing Office, February.

— (1998b) Statement of the Secretary of Agriculture at the Hearings before the House Committee on Agriculture (on the review of the 1999 WTO Multilateral Negotiations on Agricultural Trade), US House of Representatives, Washington, DC: US Government Printing Office, March.

Goldschmidt, W. (1947) *As You Sow: Three Studies in the Social Consequence of Agribusiness*, Montclair, NJ: Allanheld, Osmun and Co.

Goldsmith, E. (2004) 'Percy Schmeiser: the man that took on Monsanto', *Ecologist*, 1 May, <www.theecologist.org/archive_detail.asp?content_id=279>.

Goodland, R. (1997) 'Environmental sustainability in agriculture: diet matters', *Ecological Economics*, 23(3).

GRAIN and RAFI (Rural Advancement Foundation International) (1996) 'CGIAR: agricultural research for whom?', *Ecologist*, 26(6).

Grumbine, R. E. (1992) *Ghost Bears: Exploring the Biodiversity Crisis*, Washington, DC: Island Press.

Guha, R. (1989) 'Radical environmentalism and wilderness preservation: a Third-World critique', *Environmental Ethics*, 11(1).

Hahamovitch, C. and R. Halpern (2004) 'Not a "sack of potatoes": why labor historians need to take agriculture seriously', *International Labour and Working-class History*, 65.

Halweil, B. (2002) 'Home grown: the case for local food in a global market', *WorldWatch Paper*, 163, Washington, DC.

Harvey, D. (2003) *The New Imperialism*, New York: Oxford University Press.

Hecht, S. B. and A. Cockburn (1989) *The Fate of the Forest: Developers, Destroyers, and Defenders of the Amazon*, London: Verso.

Hedley, A. A., C. L. Ogden, C. L. Johnson, M. D. Carroll, L. R. Curtin and K. M. Flegal (2004) 'Prevalence of overweight and obesity among US children, adolescents, and adults, 1999–2002', *Journal of the American Medical Association*, 291(23).

Heffernan, W. (2000) 'Concentration of ownership and control in agriculture', in F. Magdoff, J. B. Foster and F. H. Buttel (eds), *Hungry for Profit: The Agribusiness Threat to Farmers, Food, and the Environment*, New York: Monthly Review Press, pp. 61–76.

Heffernan, W., M. Hendrickson and R. Gronski (1999) *Consolidation in the Food and Agriculture System*, Report to the National Farmers Union, Washington, DC.

Heffernan, W. D. and D. H. Constance (1994) 'Transnational corporations and the globalization of the food system', in A. Bonanno, L. Busch, W. H. Friedland, L. Gouveia and E. Mingione (eds), *From Columbus to ConAgra: The Globalization of Agriculture and Food*, Lawrence: University of Kansas Press, pp. 29–51.

Heinberg, R. (2005) *The Party's Over: Oil, War and the Fate of Industrial Societies*, Gabriola Island, BC: New Society Publishers.

Hendrickson, M. and W. Heffernan (2002) 'Concentration of agricultural markets', Report to the National Farmers Union, Washington, DC.

— (2005) 'Concentration of agricultural markets', Report to the National Farmers Union, Washington, DC.

Hendrickson, M., W. Heffernan, P. H. Howard and J. B. Heffernan (2001) 'Consolidation in food retailing and dairy', Report to the National Farmers Union, Washington, DC.

Hill, C. (1972) *The World Turned Upside Down: Radical Ideas during the English Revolution*, New York: Viking Press.

Hinton, W. (1997) *Fanshen: A Documentary of Revolution in a Chinese Village*, Berkeley: University of California Press.

— (2000) 'The importance of land reform in the reconstruction of China', in F. Magdoff, J. B. Foster and F. H. Buttel (eds), *Hungry for Profit: The*

Agribusiness Threat to Farmers, Food, and the Environment, New York: Monthly Review Press, pp. 215–29.

Hobsbawm, E. J. (1994) *The Age of Extremes: A History of the World, 1914–1991*, New York: Vintage.

Hoogvelt, A. (1997) *Globalization and the Post-Colonial World: The New Political Economy of Development*, Baltimore, MD: Johns Hopkins University Press.

Horowitz, M. H. (1971) 'Introductory essay', in M. H. Horowitz (ed.), *Peoples and Cultures of the Caribbean: An Anthropological Reader*, New York: Natural History Press, pp. 1–13.

Huizer, G. (1973) *Peasant Rebellion in Latin America*, Harmondsworth: Penguin.

— (1996) 'Social movements in the underdevelopment of development dialectic: a view from below', in S. C. Chew and R. A. Denemark (eds), *The Underdevelopment of Development: Essays in Honour of Andre Gunder Frank*, London: Sage, pp. 281–313.

Human Rights Watch (2005) *Blood, Sweat, and Fear: Workers' Rights in US Meat and Poultry Plants*, New York: Human Rights Watch.

IATP (Institute for Agriculture and Trade Policy) (2006) 'Food without thought', Report by the Institute for Agriculture and Trade Policy, Minneapolis, <www.iatp.org/iatp/publications.cfm?accountID=421&refID=80627>.

Indian Express (2006) 'One lakh farmers ended life, debt major factor: Pawar', *Indian Express*, 19 May.

IPCC (Intergovernmental Panel on Climate Change) (2001) *Climate Change 2001: Synthesis Report. Intergovernmental Panel on Climate Change*, <www.grida.no/climate/ipcc_tar/vol4/english/>.

— (2007) *Climate Change 2007: The Physical Basis,* <www.ipcc.ch/SPM2feb07.pdf>.

Jawara, F. and A. Kwa (2003) *Behind the Scenes at the WTO: The Real World of International Trade Negotiations*, London: Zed Books.

Johnson, D. G. (1998) 'Agricultural trade: future issues', in T. Yildirim, A. Schmitz and W. H. Furtan (eds), *World Agricultural Trade*, Boulder, CO: Westview, pp. 7–23.

Kay, C. (2002) 'Chile's neoliberal agrarian transformation and the peasantry', *Journal of Agrarian Change*, 2(4).

Khor, M. (2002) *Rethinking Globalization: Critical Issues and Policy Choices*, London: Zed Books.

— (2003) 'Statement by the TWN on the events of the final days of the Cancún Conference', *Third World Network*, 14 September, <www.twnside.org.sg/title/twninfo77.htm>.

— (2006) 'Impasse at the WTO: a development perspective', *Economic and Political Weekly*, 11 November.

Klein, N. (1999) *No Logo: Taking Aim at the Brand Bullies*, New York: Picador.

Kloppenburg, J. R. (2004) *First the Seed: The Political Economy of Plant Biotechnology*, 2nd edn, Madison: University of Wisconsin Press.

Kneen, B. (1995) *Invisible Giant: Cargill and Its Transnational Strategies*, Halifax: Fernwood.

Kodras, J. (1993) 'Shifting global strategies of US foreign food aid, 1955–1990', *Political Geography*, 12(3).

Kothari, A., S. Suri and N. Singh (1995) 'People and parks: rethinking conservation in India', *Ecologist*, 25(5).

Krebs, A. V. (1992) *The Corporate Reapers*, Washington, DC: Essential Books.

Kropotkin, P. (1975 [1898]) *Fields, Factories and Workshops Tomorrow*, New York: Harper and Row.

Ladd, A. E. and B. Edward (2002) 'Corporate swine and capitalist pigs: a decade of environmental injustice and protest in North Carolina', *Social Justice*, 29(3).

LaFaber, W. (1983) *Inevitable Revolutions: The US in Central America*, New York: W. W. Norton.

Lang, T. and M. Heasman (2004) *Food Wars: The Global Battle for Mouths, Minds, and Markets*, London: Earthscan.

Leakey, R. and R. Lewin (1995) *The Sixth Extinction: Patterns of Life and the Future of Humankind*, New York: Doubleday.

Lehman, K. and A. Krebs (1996) 'Control of the world's food supply', in J. Mander and E. Goldsmith (eds), *The Case against the Global Economy, and for a Turn toward the Local*, San Francisco, CA: Sierra Club Books, pp. 121–30.

Lenin, V. I. (1990 [1917]) *Imperialism: The Highest Stage of Capitalism*, New York: International Publishers.

Leonard, R. and M. A. Manahan (2004) *The Struggle for Land: A Summary of Discussions and Strategies at the Asia Land Meeting*, Bangkok and São Paulo: Thailand Land Reform Network (TLRN), Focus on the Global South and the Land Research Action Network.

Lewontin, R. C. (2000) 'The maturing of capitalist agriculture: farmer as proletarian', in F. Magdoff, J. B. Foster and F. H. Buttel (eds), *Hungry for Profit: The Agribusiness Threat to Farmers, Food, and the Environment*, New York: Monthly Review Press, pp. 93–106.

Leys, C. (1996) *The Rise and Fall of Development Theory*, London: James Currey.

Lilliston, B. and R. Cummins (1998) 'Organic vs. "organic": the corruption of a label', *Ecologist*, 28(4).

Lipton, M. (1977) *Why Poor People Stay Poor: Urban Bias in World Development*, Cambridge, MA: Harvard University Press.

Little, P. D. and M. J. Watts (1994) *Living under Contract: Contract Farming and Agrarian Transformation in Sub-Saharan Africa*, Madison: University of Wisconsin Press.

Lohmann, L. (1993) 'Against the myths', in M. Colchester and L. Lohmann (eds), *The Struggle for Land and the Fate of the Forests*, London: Zed Books.

Lovell, W. G. (1995) *A Beauty that Hurts: Life and Death in Guatemala*, Toronto: Between the Lines.

Lovelock, J. (1987) *Gaia: A New Look at Life on Earth*, New York: Oxford University Press.

Lutzenberger, J. and M. Halloway (1999) 'The absurdity of modern agriculture: from chemical fertilizers and apropoisons to biotechnology', in G. Tansey and J. D'Silva (eds), *The Meat Business: Devouring a Hungry Planet*, New York: St Martin's Press, pp. 3–14.

Lynas, M. (2004) 'If they plant them, we'll pull them up', *Ecologist*, 34(3).

Lyson, T. A. and A. Lewis Raymer (2000) 'Stalking the wily multinational: power and control in the US food system', *Agriculture and Human Values*, 17(2).

McCully, P. (1996) *Silenced Rivers: The Ecology and Politics of Large Dams*, London: Zed Books.

McMichael, P. (2000a) 'The power of food', *Agriculture and Human Values*, 17(1).

— (2000b) 'Global food politics', in F. Magdoff, J. B. Foster and F. H. Buttel (eds), *Hungry for Profit: The Agribusiness Threat to Farmers, Food, and the Environment*, New York: Monthly Review Press, pp. 125–43.

— (2004a) 'Biotechnology and food security: profiting on insecurity', in L. Beneria and S. Bisnath (eds), *Global Tensions: Challenges and Opportunities in the World Economy*, New York: Routledge.

— (2004b) 'Global development and the corporate food regime', Paper prepared for the Symposium on New Directions in the Sociology of Global Development, XI World Congress of Rural Sociology, Trondheim, July.

McNally, D. (2002) *Another World is Possible: Globalization and Anti-Capitalism*, Winnipeg: Arbeiter Ring.

Magdoff, F. (2004) 'A precarious existence: the fate of billions?', *Monthly Review*, 55(9).

Magdoff, F., J. B. Foster and F. H. Buttel (2000) 'An overview', in F. Magdoff, J. B. Foster and F. H. Buttel (eds), *Hungry for Profit: The Agribusiness Threat to Farmers, Food, and the Environment*, New York: Monthly Review Press, pp. 7–21.

Majka, L. C. and T. J. Majka (2000) 'Organizing US farm workers: a continuous struggle', in F. Magdoff, J. B. Foster and F. H. Buttel (eds), *Hungry for Profit: The Agribusiness Threat to Farmers, Food, and the Environment*, New York: Monthly Review Press, pp. 161–74.

Mallin, M. (2000) 'Impacts of industrial animal agriculture on rivers and estuaries', *American Scientist*, 88(1).

Marks, R. (2001) *Cesspools of Shame: How Factory Farm Lagoons and Sprayfields Threaten Environmental and Public Health*, Washington, DC: Natural Resources Defense Council and the Clean Water Network, <www.nrdc.org/water/pollution/cesspools/cesspools.pdf>.

Mattera, P. (2004) *USDA Inc.: How Agribusiness Has Hijacked Regulatory Policy at the US Department of Agriculture*, Chicago, IL: Agribusiness Accountability Initiative and Corporate Research Project, <www.agribusinessaccountability.org/page/325/1>.

Mason, J. and P. Singer (1990) *Animal Factories*, 2nd edn, New York: Harmony Books.

Masson, J. M. (2003) *The Pig Who Sang to the Moon: The Emotional World of Farm Animals*, New York: Ballantine.

Mazoyer, M. and L. Roudart (2006) *A History of World Agriculture: From the Neolithic Age to the Current Crisis*, New York: Monthly Review Press.

Mellon, M., C. Benbrook and K. Lutz Benbrook (2001) 'Hogging it! Estimates of antimicrobial abuse in livestock', Washington, DC: Union of Concerned Scientists.

Melville, E. (1994) *A Plague of Sheep: Environmental Consequences of the Conquest of Mexico*, Cambridge: Cambridge University Press.

Mendis, A. and C. Van Bers (1999) 'Bitter fruit: attractive supermarket displays of tropical fruit conceal ugly environmental and social costs', *Alternatives Journal*, 25(1).

Middendorf, G., M. Skladny, E. Ransom and L. Busch (2000) 'The great global enclosure of our times: peasants and the agrarian question at the end of the twentieth century', in F. Magdoff, J. B. Foster and F. H. Buttel (eds), *Hungry for Profit: The Agribusiness Threat to Farmers, Food, and the Environment*, New York: Monthly Review Press, pp. 107–23.

Midkiff, K. (2004) *The Meat You Eat: How Corporate Farming Has Endangered America's Food Supply*, New York: St Martin's Press.

Mintz, S. W. (1985) *Sweetness and Power: The Place of Sugar in Modern History*, New York: Viking.

Mittal, A. (2003) 'Robin Hood in reverse', *New Internationalist*, 353.

Morgan, D. (1980) *Merchants of Grain*, New York: Penguin.

MST (Movimento dos Trabalhadores Rurais Sem Terra) Secretariat (2006) 'The perverse nature of agribusiness for Brazilian society', *ZNet*, 22 February, <www.zmag.org/content/print_article.cfm?itemID=9786& sectionID=48>.

Murray, D. L. (1995) *Cultivating Crisis: The Human Cost of Pesticides in Latin America*, Austin: University of Texas Press.

Myers, N. (1981) 'The hamburger connection: how Central America's forests become North America's hamburgers', *Ambio*, 10(1).

Nations, J. D. and D. I. Komer (1987) 'Rainforests and the hamburger society', *Ecologist*, 17(4).

Nestle, M. (2002) *Food Politics: How the Food Industry Influences Nutrition and Health*, Berkeley: University of California Press.

Netting, R. M. (1993) *Smallholders, Householders: Farm Families and the Ecology of Intensive, Sustainable Agriculture*, Stanford, CA: Stanford University Press.

Nierenberg, D. (2003) 'Factory farming in the developing world', *WorldWatch Magazine*, 16(3).

— (2005) *Happier Meals: Rethinking the Global Meat Industry*, WorldWatch Paper no. 171, Washington, DC.

Nord, M., M. Andrews and S. Carlson (2005) *Household Food Security in the United States, 2004*, Economic Research Report no. ERR11, Washington, DC: United States Department of Agriculture, Economic Research Service.

Novacek, M. J. (2001) *Biodiversity Crisis: Losing What Counts*, New York: New Press.

NYT (New York Times) (2003) 'Trade rigged against the poor', 21 July.

— (2006) 'Nestlé set to purchase diet firm Jenny Craig', 19 June.

OECD (Organization for Economic Cooperation and Development) (1999) *OECD Agricultural Outlook 1999–2004*, Paris: OECD.

Office of the United Nations High Commission for Human Rights (2004) 'The submission of the Special Rapporteur on the right to the General Assembly as requested by its resolution 58/186 and Commission on Human Rights resolution 2004/19', <www.ohchr.org/english/bodies/chr/docs/ga59/newfood.doc>.

Paarlberg, R. L. (2000) 'The global food fight', *Foreign Affairs*, 79(3).

Paoletti, M. G., D. Pimentel, B. R. Stinner and D. Stinner (1992) 'Agroecosystem biodiversity: matching production and conservation biology', *Agriculture, Ecosystems and Environment*, 40(1–4).

Petras, J. and H. Veltmeyer (2001) *Globalization Unmasked: Imperialism in the 21st Century*, London: Zed Books.

Pimentel, D. (2005) 'Environmental and economic costs of the application of pesticides primarily in the United States', *Environment, Development and Sustainability*, 7(2).

Pimentel, D. and H. Lehman (eds) (1993) *The Pesticide Question: Environment, Economics and Ethics*, New York: Chapman and Hall.

Pinstrup-Andersen, P. (2000) 'Food policy research for developing countries: emerging issues and unfinished business', *Food Policy*, 25(2).

Plant, R. (1993) 'Background to agrarian reform: Latin America, Asia, and Africa', in M. Colchester and L. Lohmann (eds), *The Struggle for Land and the Fate of the Forests*, London: Zed Books.

Pretty, J. N. (1995) *Regenerating Agriculture: Policies and Practice for Sustainability and Self-reliance*, London: Earthscan.

— (1997) 'The sustainable intensification of agriculture', *Natural Resources Forum*, 21(4).

Ramakrishnan, V. (2006) 'Seeds and protests', *Frontline*, 23(12).

Rampersad, F., T. Stewart, G. Ramprasad and R. Ramprasad (1997) *The New World Order: Uruguay Round Agreements and Implications for CARICOM States*, Kingston: Ian Randle and UWI Institute of Social and Economic Research.

Reardon, T., C. P. Timmer, C. B. Barrett and J. Berdegue (2003) 'The rise of supermarkets in Africa, Asia and Latin America', *American Journal of Agricultural Economics*, 85(5).

Rees, W. E. and L. Westra (2003) 'When consumption does violence: can there be sustainability and environmental justice in a resource-limited world?', in J. Agyeman, R. D. Bullard and B. Evans (eds), *Just Sustainabilities: Development in an Unequal World*, Cambridge, MA: MIT Press, pp. 99–124.

Regan, T. (2004) *The Case for Animal Rights*, Berkeley: University of California Press.

Reisner, M. (1993) *Cadillac Desert: The American West and Its Disappearing*

Water, 2nd edn, New York: Penguin.

Ridgeway, J. (2004) *It's All for Sale: The Control of Global Resources*, Durham, NC: Duke University Press.

Rifkin, J. (1992) *Beyond Beef: The Rise and Fall of the Cattle Culture*, New York: Penguin.

Robbins, P. (2003) *Stolen Fruit: The Tropical Commodities Disaster*, London: Zed Books.

Rodney, W. (1972) *How Europe Underdeveloped Africa*, Washington, DC: Howard University Press.

Ross, J. (2005) 'Paying the price for growth: fruit, vegetable production spurs Chile's exports', *Toronto Star*, 8 January.

Rosset, P. M. (1999a) *The Multiple Functions and Benefits of Small Farm Agriculture in the Context of Global Trade Negotiations*, Food First Policy Brief no. 4, San Francisco: Institute for Food and Development Studies, September.

— (1999b) 'Small is bountiful', *Ecologist*, 29(8).

— (2000) 'Cuba: a successful case study of alternative agriculture', in F. Magdoff, J. B. Foster and F. H. Buttel (eds), *Hungry for Profit: The Agribusiness Threat to Farmers, Food, and the Environment*, New York: Monthly Review Press, pp. 203–13.

— (2006) *Food is Different: Why We Must Get the WTO Out of Agriculture*, London: Zed Books.

Roy, A. (1999) 'The greater common good', *Frontline*, 16(11).

Sachs, W. (1999) *Planet Dialectics: Explorations in Environment and Development*, London: Zed Books.

Sanjay, S. J. and T. W. Pogge (2005) 'How *not* to count the poor', Colombia University Technical Paper, Version 6.2, 29 October, <www.columbia.edu/~sr793/count.pdf>.

SAPRIN (Structural Adjustment Participatory Review International Network) (2004) *Structural Adjustment: The Policy Roots of Economic Crisis, Poverty, and Inequality*, London: Zed Books.

Sauer, C. O. (1952) *Agricultural Origins and Dispersals*, New York: American Geographical Society.

— (1972 [1952]) *Seeds, Spades, Hearths and Herds: The Domestication of Animals and Foodstuffs*, Cambridge, MA: MIT Press.

Scher, P. L. (1998) Testimony of the Special Trade Negotiator for Agriculture before the House Committee on Agriculture (on the review of the 1999 WTO Multilateral Negotiations on Agricultural Trade), US House of Representatives, Washington, DC: US Government Printing Office, 18 March.

Schlesinger, S. C. and S. Kinzer (1982) *Bitter Fruit: The Untold Story of the American Coup in Guatemala*, Garden City, NY: Doubleday.

Schlosser, E. (2002) *Fast Food Nation: The Dark Side of the All-American Meal*, New York: Perennial.

Schmeiser, P. (2004) 'Theft of life: a story of the struggle against Monsanto and the corporate takeover of our genetic inheritance', *Resurgence*, 223, March/April.

Bibliography

Schumacher, A., Jr (1998) Statement of the USDA Under Secretary for Farm and Foreign Agricultural Services before the Subcommittee on Trade, Committee on Ways and Means, US House of Representatives, Washington, DC: US Government Printing Office, 12 February.

Sen, A. (1989) 'Food and freedom', *World Development*, 17(6).

Sexton, S. (1996) 'Transnational corporations and food', *Ecologist*, 26(6).

Shiva, V. (1991) *The Violence of the Green Revolution: Third World Agriculture, Ecology, and Politics*, London: Zed Books.

— (1993) *Monocultures of the Mind: Perspectives on Biodiversity and Biotechnology*, London: Zed Books.

— (1999) 'Monocultures, monopolies, myths and the masculinization of agriculture', *Development: Journal of the Society for International Development*, 42(2).

Silverstein, K. (1999) 'Meat factories: hellish hog plants, lakes of sewage, and lifeless waterways – are cheap bacon-burgers worth it?', *Sierra*, January/February.

Simon, J. (1981) *The Ultimate Resource*, Princeton, NJ: Princeton University Press.

Sinclair, U. (1981 [1906]) *The Jungle*, New York: Bantam.

Smil, V. (2001) *Enriching the Earth: Fritz Haber, Carl Bosch, and the Transformation of World Food Production*, Cambridge, MA: MIT Press.

Smith, N. J. H. and D. Forno (1996) *Biodiversity and Agriculture: Implications for Conservation and Development*, Technical Paper no. 321, Washington, DC: World Bank.

Stedile, J. P. (2007) 'The neoliberal agrarian model in Brazil', *Monthly Review*, 58(9).

Stern, N. (2006) *Stern Review on the Economics of Climate Change*, <www.hm-treasury.gov.uk/independent_reviews/stern_review_economics_climate_change/stern_review_report.cfm>.

Stiglitz, J. E. (2003) *Globalization and Its Discontents*, New York: W. W. Norton.

Stull, D., M. Broadway and D. Griffith (1995) *Any Way You Cut It: Meat Processing and Small-town America*, Lawrence: University of Kansas Press.

Sun, L. H. and G. Escobar (1999) 'On chicken's front line', *Washington Post*, 28 November.

Suri, K. C. (2006) 'Political economy of agrarian distress', *Economic and Political Weekly*, 22 April.

Swaminathan, M. S. (2006) 'Wheat imports and food security', *Frontline*, 23(12).

Swift, M. J. and J. M. Anderson (1993) 'Biodiversity and ecosystem function in agricultural systems', in E.-D. Schulze and H. A. Mooney (eds), *Biodiversity and Ecosystem Function*, New York: Springer-Verlag.

Thiesenhusen, W. (1991) 'Implications of the rural land tenure system for the environmental debate: three scenarios', *Journal of Developing Areas*, 26(10).

Thrupp, L. A. with G. Bergeron and W. F. Waters (1995) *Bittersweet Harvests for Global Supermarkets: Challenges in Latin America's Agricultural Export Boom*, Washington, DC: World Resources Institute.

Tilman, D., K. G. Cassman, P. A. Matson, R. Naylor and S. Polasky (2002) 'Agricultural sustainability and intensive production practices', *Nature*, 418.

Time (2006) 'Inside the pitchfork rebellion', 13 March.

Tolchin, T. (1998) 'Wasting away: big agribusiness factory farms make a big mess', *Multinational Monitor*, 19(6).

Trouillot, M.-R. (1994) 'Haiti's nightmare and the lessons of history', *NACLA Report on the Americas*, 27(4).

Tweeten, L. G. (1997) 'Food security', in L. G. Tweeten and D. G. McClelland (eds), *Promoting Third World Development and Food Security*, Westport, CT: Praeger, pp. 225–56.

UNDP (United Nations Development Programme) (1998) *Human Development Report*, New York: Oxford University Press.

UNHSP (United Nations Human Settlements Programme) (2003) *The Challenge of Slums: Global Report on Human Settlements 2003*, London: Earthscan.

USCB (United States Census Bureau) (1999) *The National Data Book: Statistical Abstract on the US 1999*, 119th edn, Lanham, MD: Bernan Press.

Veltmeyer, H. (2005) 'The dynamics of land occupation in Latin America', in S. Moyo and P. Yeros (eds), *Reclaiming the Land: The Resurgence of Rural Movements in Africa, Asia and Latin America*, London: Zed Books, pp. 286–316.

Vidal, J. (2004) 'Demand for beef speeds destruction of Amazon forest', *Guardian Weekly*, 2 April.

Von Braun, J. and M. A. Brown (2003) 'Ethical questions of equitable worldwide food production systems', *Plant Physiology*, 133(11).

Wackernagel, M. and W. E. Rees (1996) *Our Ecological Footprint: Reducing Human Impact on Earth*, Gabriola Island, BC: New Society Publishers.

Walker, A. (1988) *Living by the Word: Selected Writings 1973–1986*, New York: Harcourt Brace Jovanovich.

Wallerstein, I. (1997) 'Ecology and capitalist costs of production: no exit', Address given at PEWS XXI: The Global Environment and the World-System, University of California, Santa Cruz, 3–5 April.

Walton, J. and D. Seddon (1994) *Free Markets and Food Riots: The Politics of Global Adjustment*, Oxford: Blackwell.

Watkins, K. (1996) 'Free trade and farm fallacies: from the Uruguay Round to the World Food Summit', *Ecologist*, 26(6).

— (2003) 'The real Cancún', *Australian Financial Review*, 22 August.

Watts, M. J. (1983) *Silent Violence: Food, Famine and Peasantry in Northern Nigeria*, Berkeley: University of California Press.

— (2005) 'Commodities', in P. Cloke, P. Crang and M. Goodwin (eds), *Introducing Human Geographies*, 2nd edn, London: Edward Arnold, pp. 527–46.

Weiner, T. (1999) 'It's raining farm subsidies', *New York Times*, 8 August.

Weis, T. (2003) 'Agrarian decline and breadbasket dependence in the Caribbean: confronting illusions of inevitability', *Labour, Capital, and Society*, 36(2).

— (2004a) 'Restructuring and redundancy: the impact and illogic of neoliberal agricultural reforms in Jamaica', *Journal of Agrarian Change*, 4(4).

— (2004b) '(Re-)making the case for land reform in Jamaica', *Social and Economic Studies*, 53(1).

— (2006) 'The rise, fall, and future of the Jamaican peasantry', *Journal of Peasant Studies*, 33(1).

WHO (World Health Organization) (1992) *Our Planet, Our Health: Report of the WHO Commission on Health and Environment*, Geneva: World Health Organization.

WHO and UNEP (United Nations Environmental Programme) (1990) *Public Health Implications of Pesticides Used in Agriculture*, Geneva: World Health Organization.

Williamson, J. (1990) 'What Washington means by policy reform', in J. Williamson (ed), *Latin American Adjustment: How Much Has Happened*, Washington, DC: Institute for International Economics, pp. 5–20.

Wilson, E. O. (2002) *The Future of Life*, New York: Vintage Books.

Winters, L. A. (1988) 'The so-called "non-economic" objectives of agricultural policy', Working Paper no. 52, Department of Economics and Statistics, Paris: OECD.

Wolf, E. R. (1969) *Peasant Wars of the Twentieth Century*, New York: Harper and Row.

Wood, D. (1996) 'The benign effect of some agricultural specialization on the environment', *Ecological Economics*, 19(2).

Wood, E. M. (2000) 'The agrarian origins of capitalism', in F. Magdoff, J. B. Foster and F. H. Buttel (eds), *Hungry for Profit: The Agribusiness Threat to Farmers, Food, and the Environment*, New York: Monthly Review Press, pp. 23–42.

— (2002) *The Origin of Capitalism: A Longer View*, London: Verso.

World Bank (2001) *Facets of Globalization: International and Local Dimensions of Development*, Washington, DC: World Bank.

— (2002) *Global Economic Prospects and the Developing Countries 2002: Making Trade Work for the World's Poor*, Washington, DC: World Bank.

World Commission on Environment and Development (1987) *Our Common Future*, New York: Oxford University Press.

WorldWatch (2004) 'Meat: now, it's not personal! But like it or not, meat-eating is becoming a problem for everyone on the planet', *WorldWatch Magazine*, July/August.

Worster, D. (1993) *The Wealth of Nature: Environmental History and the Ecological Imagination*, New York: Oxford University Press.

Yates, M. D. (2004) 'Poverty and inequality in the global economy', *Monthly Review*, 55(9).

Ziegler, J. (2004) 'The right to food', Report of the Special Rapporteur of the United Nations Commission on Human Rights, submitted to the General Assembly, New York.

Zola, E. (1980 [1887]) *The Earth*, London: Penguin.

Index

Index